BURLEIGH DODDS SCIENCE: INSTANT INSIGHTS

NUMBER 14

Bone health in poultry

Published by Burleigh Dodds Science Publishing Limited
82 High Street, Sawston, Cambridge CB22 3HJ, UK
www.bdspublishing.com

Burleigh Dodds Science Publishing, 1518 Walnut Street, Suite 900, Philadelphia, PA 19102-3406, USA

First published 2021 by Burleigh Dodds Science Publishing Limited

British Library Cataloguing in Publication Data
A catalogue record for this book is available from the British Library

ISBN 978-1-80146-012-5 (Print)
ISBN 978-1-80146-013-2 (ePub)

DOI 10.19103/9781801460132

Typeset by Deanta Global Publishing Services, Dublin, Ireland

Contents

Chapter 1

Genetics and genomics of skeletal traits in poultry species

Martin Johnsson, Swedish University of Agricultural Sciences, Sweden

1 Introduction

Impaired skeletal function in poultry occurs both in meat-type and layer birds, with different aetiologies (see reviews by Pines and Reshef (2015) and Toscano (2018) for layer chickens and Bradshaw et al. (2002) for broiler chickens, specifically). These problems include keel bone fractures, osteoporosis, long bone deformities and poor bone mineralisation. While environmental, dietary and management factors play key roles in skeletal traits, there are also substantial genetic influences, and they are the topic of this chapter.

Key research questions in this area include: What are the quantitative genetic properties of skeletal traits, such as heritabilities and genetic correlations? What genes and genetic variants contribute to variation in skeletal traits? How should we measure bone traits? After a brief overview of bone biology, this chapter reviews these questions in turn, and ends with possible future directions.

2 Avian bone biology

Bone pathologies can potentially derive from faulty bone development or faulty bone homeostasis. The genetic regulation of bone patterning and

http://dx.doi.org/10.19103/AS.2020.0065.16

shape and the differentiation of bone cells (chondrocytes, osteoblast and osteoclasts) are molecularly distinct (Karsenty and Wagner, 2002). Thus, the bone problems common among meat-type chickens (bone deformations and dyschondroplasia) and the ones common among layers (osteoporosis) are likely to have independent genetic influences. Furthermore, genetic improvement in bone traits could come about through effects on any of the cell types involved at different developmental times.

Long bones develop by endochondral ossification, where the cartilage is first formed followed by mineralisation. Later, long bones grow by cartilage ossification at the growth plate at the end of the bone (metaphysis). Dyschondroplasia occurs when the cartilage at the end of bones, such as the tibiotarsus, fails to mature and ossify (Farquharson and Jefferies, 2000). Instead, it forms abnormal cartilage masses at the metaphysis. It is one of the major leg problems in broiler chickens. Broiler chickens also have problems with long bone deformation, where bones are angled away from the body (valgus or 'bow-legged') or towards the body (varus or 'knock-kneed'), rotated or twisted.

The main problem for layer chickens is poor bone quality. After growth, bone is in a constant state of remodelling through an interplay of bone cells of different origins. Osteoblasts, derived from mesenchymal stem cells, form bone. Osteoclasts, derived from macrophages, break down bone. For healthy bone, these two processes should be in a balance to allow growth or maintenance of bone in the face of challenges such as sustained egg production. Disturbance of this bone homeostasis can cause osteoporosis, with resulting fractures and bone deformations.

The presence of medullary bone is a key difference between avian and mammalian bone. Bone is made up of an outer cortical part, which is denser, and an inner, trabecular part, which is less dense. In female birds, there is a third type of bone: the medullary bone. The medullary cavity contains medullary bone, which is similar to trabecular bone in its lower density. It remodels rapidly and serves as a source and store of calcium for eggshell production (Dacke et al., 1993; Van de Velde et al., 1984). At sexual maturity, cortical and trabecular bone stop forming and medullary bone is formed (Bloom et al., 1958).

During lay, the medullary bone turns over rapidly to provide calcium for eggshell formation. The cortical bone that provides more structural support decreases. Thus, bones gradually weaken during the egg laying period (Cransberg et al., 2001; Wilson et al., 1992) as a physiological consequence of sustained egg production. The keel bone, the enlarged avian breastbone or sternum, is particularly exposed to deformities and fractures. Keel bone fractures and deformities is one of the major bone problems observed in laying hens (see review by Riber et al. (2018)).

In summary, skeletal problems in broilers and layers stem from different failures of bone development and maintenance, which are physiologically related to growth and egg production, respectively. It is the task of quantitative genetics to discover the extent to which breeding can alleviate these problems, and the extent that these physiological relationships translate to genetic correlations.

3 Quantitative genetics of bone traits

There have been several kinds of quantitative genetic studies on bone traits:

1. Selection experiments demonstrate that bone traits can respond to selection, and the potential correlated selection responses.
2. Studies where different lines of chickens are raised in uniform conditions demonstrate genetic differences.
3. Pedigree-based quantitative genetic studies allow the estimation of genetic variances and covariances.
4. Retrospective evaluation of breeding practices shows how selection has improved bone traits.

3.1 Selection experiments

Bishop et al. (Bishop et al., 2000) performed a selection experiment on bone strength in layers, selecting on an index that involved tibial and humeral breaking strength, keel radiographic density and body weight, intending to change the bone traits without changing body weight. Bone traits responded to selection, with a concomitant difference in the number of fractures, but little or no change in egg laying rate or egg quality. Selection was retrospective, based on bone parameters measured at the end of the laying period. This study demonstrated that selection could improve bone strength in layers. Follow-up studies gave indications about the mechanisms and consequences of the differences:

- The high bone-strength line had more pyrrole crosslinks in their bone collagen matrix (Sparke et al., 2002), while the amounts of collagen and calcium did not differ much. Pyrrole crosslinks are also affected in bones of osteoporotic hens (Knott et al., 1995).
- However, a later study found that the lines had similar bone material composition in terms of both minerals and collagen (Rodriguez-Navarro et al., 2018).
- The high bone-strength line had higher keel-bone density and less keel-bone deformities (Fleming et al., 2004; Stratmann et al., 2016).
- The high bone-strength line had fewer osteoclasts and more of the medullary bone, indicative of lower bone-resorption rate (Fleming et al., 2006).
- Egg-shell quality was not different between the lines (Fleming et al., 2006).

When it comes to broiler chickens, several selection studies showed how incidence of tibial dyschondroplasia responded favourably to selection (Riddell, 1976; Wong-Valle et al., 1993; Zhang et al., 1998). These results showed that tibial dyschondroplasia could be reduced by selection, but there were conflicting results about unfavourable responses in production-related traits (Yalcin et al., 2000; Zhang et al., 1995).

3.2 Line comparisons

Hocking et al. (Hocking et al., 2003, 2009) demonstrated heritable differences in bone and egg parameters in various layer, broiler and traditional chicken breeds by rearing them in a uniform environment. Commercial layer chickens had weaker bones than traditional breeds, but comparable egg-shell strength, and similar bone size and morphology. Specifically, the commercial chickens were losing structural bone, but maintaining medullary bone, which is consistent with a trade-off between bone quality and laying of large numbers of high-quality eggs.

Kestin, Su and Sorensen (Kestin et al., 1999) found differences in the propensity for leg problems between different broiler-line crosses. Four commercial crossbreds differed in the number of leg defects and the birds' walking abilities, consistent with genetic differences.

3.3 Heritability estimates

There have been many estimates of the heritability of bone traits, particularly in broiler chickens (Table 1). In general, bone traits are weakly to moderately heritable, with estimates for skeletal defects ranging from 0.04 to 0.3 and estimates for bone breaking strength ranging from 0.2 to 0.5. In addition to genetic differences between populations and the possibility that visually scored traits were scored differently, it bears mentioning that quantitative genetic estimation of parameters related to categorical traits is not straightforward, and the studies underlying Table 1 sometimes took different approaches.

3.4 Genetic correlations

There have been many estimates of correlations between skeletal problems of body weights in broilers, but less published work on genetic correlations between bone quality and production traits in layers. Table 2 summarises the estimates of genetic correlation between bone traits and production traits from the literature.

In summary, the genetic correlations between bone defects and growth in broilers were variable, but in general weakly positive (i.e. unfavourable). This

Table 1 Heritability estimates of bone traits

Citation	Trait	Heritability estimates	Chickens
Tibial dyschondroplasia			
Kuhlers and McDaniel (1996)	Tibial dyschondroplasia (low-intensity X-ray imaging)	0.37 (4 weeks), 0.42 (7 weeks)	Broiler
Kapell et al. (2012)	Tibial dyschondroplasia (low-intensity X-ray imaging)	0.27, 0.10, 0.15	Broiler
Rekaya et al. (2013)	Tibial dyschondroplasia (post-mortem scoring)	0.12	Broiler
Siegel et al. (2019)	Tibial dyschondroplasia	0.13, 0.16, 0.18	Broiler
Other defects			
Mercer and Hill (1984)	Leg problems	0.09, 0.21, 0.15 [1]	Broiler
	Keel problems	0.06, 0.11, 0.09 [1]	
Le Bihan-Duval et al. (1996)	Valgus	0.16, 0.29 [2]	Broiler
	Varus	0.21, 0.24 [2]	
Kapell et al. (2012)	Long bone deformity	0.068, 0.037, 0.051	Broiler
	Crooked toes	0.014, 0.10, 0.018	
Rekaya et al. (2013)	Valgus	0.23	Broiler
	Varus	0.26	
	Rotated tibia	0.18	
Siegel et al. (2019)	Valgus	0.14, 0.10, 0.11	Broiler
	Varus	0.10, 0.13, 0.14	
	Rotated legs	0.09, 0.02, 0.04	
	Femur head necrosis	0.26, 0.29, 0.30	
Bone-breaking strength and composition			
Bishop et al. (2000)	Keel-bone density (autoradiography)	0.39	Layer
	Humeral breaking strength	0.30	
	Tibial breaking strength	0.45	
González-Cerón et al. (2015a)	Tibial density	0.23	Broiler
	Tibial breaking strength	0.18	
	Tibial mineral density	0.13	
	Tibial mineral content	0.26	
	Tibial ash content	0.08	
Mignon-Grasteau et al. (2016)	Tibial ash content relative to body weight	0.19	Broiler
	Tibial breaking strength	0.45	
	Phosphorous retention	0.080	
	Plasma phosphorous	0.055	
Preisinger (2018)	Keel bone score	0.30, 0.15	Layer
	Humeral ultrasound	0.20, 0.17	

[1] Paternal half-sib heritabilities calculated with proband method, for three different lines. The composite traits 'leg problems' and 'keel problems' are made up of several defects, lumped because they were genetically correlated.
[2] Sire–grandsire heritabilities.

Table 2 Genetic correlation estimates of bone traits and production traits (when a study reported many genetic correlations, the table presents an example)

Citation	Bone trait	Production trait	Correlation	Chickens
Mercer and Hill (1984)	Leg problems	Body weight	0.24, 0.33, 0.33	Broiler
	Keel problems	Body weight	−0.14, −0.03	
Zhang et al. (1995)	Tibial dyschondroplasia	Body weight	−0.46	Broiler
Kuhlers and McDaniel (1996)	Tibial dyschondroplasia (low-intensity X-ray imaging)	Body weight	0.10 (4 weeks), 0.07 (7 weeks)	Broiler
Bihan-Duval et al. (1997)	Valgus	Body weight (6 weeks)	0.09, 0.01	Broiler
	Varus	Body weight (6 weeks)	0.08, −0.06	
Bishop et al. (2000)	Keel radiographic density	Body weight	0.36	Layer
	Humeral breaking strength	Body weight	0.26	
	Tibial breaking strength	Body weight	0.33	
Hocking et al. (2003)	Keel density (autoradiography)	Egg-shell strength	0.36 [1]	Layer
	Cortical bone content	Egg-shell strength	−0.23 [1]	
Kapell et al. (2012)	Long bone deformity	Body weight	0.089, 0.20, 0.25	Broiler
	Crooked toes	Body weight	0.15, 0.16, 0.13	
	Tibial dyschondroplasia (low-intensity X-ray imaging)	Body weight	0.19, 0.22, 0.16	
Rekaya et al. (2013)	Tibial dyschondroplasia (post-mortem scoring)	Body weight	−0.03	Broiler
	Valgus	Body weight	−0.21	
	Varus	Body weight	−0.02	
	Rotated tibia	Body weight	0.04	
González-Cerón et al. (2015a)	Tibial mineral density	Body weight gain (weeks 0–6)	0.18	Broiler
	Tibial mineral content	Body weight gain (weeks 0–6)	0.65	
	Tibial ash content	Body weight gain (weeks 0–6)	−0.09	
	Tibial breaking strength	Body weight gain (weeks 0–6)	−0.04	

[1] Correlation between breed averages.

means that in the absence of concomitant selection against them, selection for growth would likely increase defects. As will be seen below, broiler breeding programmes have practiced selection against bone defects.

Unsurprisingly, bone dimensions genetically correlate positively with body weight and growth (e.g. González-Cerón et al., 2015a), but bone-quality traits can show unfavourable correlations. That is, bones grow bigger as the body grows bigger, but with lower bone quality. There is a phenotypic correlation between bone quality and egg laying (Johnsson et al., 2016; Schreiweis et al., 2005; Whitehead, 2004), and the between-breed correlation reported by Hocking et al. (2003) suggests that that this trade-off between egg production and bone quality is also reflected in genetics. On the contrary, Fleming et al. (2006) found no difference in egg-shell quality between lines selected for bone strength. Precise quantifications of the genetic relationship between production traits and bone-quality traits in layers are yet to be published.

Different measures of skeletal defects are only weakly positively genetically correlated (Kapell et al., 2012). This suggests that, in practice, one needs to score several aspects of skeleton health. The genetic correlations between bone quality and skeletal defects are low (González-Cerón et al., 2015b), likely reflecting different mechanisms underlying skeletal defects and osteoporosis.

3.5 *Genetic trends*

The quantitative genetic results discussed above suggest that breeding against leg defects in broilers should be relatively efficient. In summary, different defect traits are somewhat heritable, their unfavourable genetic correlations with body weight appear at worst moderate, and phenotyping live birds is relatively practical using visual scoring and low-intensity X-ray imaging.

This result has been borne out in retrospective analysis of data from the Aviagen and Cobb-Vantress broiler breeding programmes. Kapell et al. (2012) report decreasing phenotypic trends for long-bone deformity, crooked toes and tibial dyschondroplasia in two lines. Siegel et al. (2019) report decreasing genetic trends for valgus, varus, rotated limbs and tibial dyschondroplasia in three lines, but no genetic improvement in femur head necrosis.

4 Genetic mapping of bone traits

Linkage mapping and genome-wide association studies have identified numerous loci associated with bone traits. Table 3 summarises genetic mapping studies of bone traits in chickens.

Genetic mapping methods localise genetic variants that affect traits. The genomic resolution is almost never fine-grained enough to isolate individual causative genes and variants. However, several studies have identified plausible candidate genes located close to association signals. For example, Zhang et al. (2011) fine-mapped one quantitative trait locus down to a handful of genes, one of which, *RB1*, is known to be involved in bone development. Guo et al. (2017) identified associations close to, among others, bone genes *RANK*,

ADAMTS15 and *SOST*. Johnsson et al. (2015) combined genetic mapping with expression QTL mapping, and found several candidates known to be involved in bone, including *COL11A1*, *osteonectin* and *HSF5*. However, in all these cases, evidence of causality is lacking, and one cannot rule out that the association may be due to linked genes. Convincing causative variants for bone traits in poultry remain to be isolated.

Crosses of the White Leghorns divergently selected for bone have been used for genetic mapping, and found a large-effect signal on chromosome 1 (Dunn et al., 2007). Further studies of this region, using further generations of the founder line, allowed fine-mapping of the association and functional genomic studies using RNA-seq (de Koning et al., 2020). This demonstrated a large difference in expression of the cystathionine beta synthase gene (*CBS*), which is at the locus, as well as differences in concentration of its substrate, homocysteine. Homocysteine is known to affect collagen crosslinking.

The published genetic mapping studies use covariate adjustments inconsistently. Given the environmental influences on bone traits, including diet, housing, laying state and the correlation with body weight, attention to covariates is important to reduce noise and also make sure that the mapping does not instead detect body weight or egg production loci. For example, when measuring bone quality-related traits:

- Zhang et al. (2011) used hatch weight for traits measured before 12 weeks of age, and carcass weight for others.
- Schreiweis et al. (2005) used cumulative egg production, body weight and for some traits also body weight squared.
- Zhou et al. (2007) appear to have used no adjustment for body weight or egg laying.

There are some indications of pleiotropic or closely linked effects on other traits. While genetic correlations estimate the overall relationship between traits across all loci, overlap of quantitative trait loci has the potential to find individual pleiotropic variants. Quantitative trait loci for bone traits also overlap associations for blood calcium (Van Goor et al., 2016), onset of sexual maturity (Wright et al., 2012), egg laying and comb size (Wright et al., 2010). The connections between bone, egg and comb traits may suggest a connection to endocrine function. As with causal gene-identification, genomic resolution is limited in distinguishing linkage from genuine pleiotropy. For example, the regions of blood calcium region identified by Van Goor et al. (2016) spanned about 1 Mbp, and contained (at the time, with Ensembl annotation and the Galgal4 chicken genome) 48 genes. The sexual maturity loci identified in Wright et al. (2012) have an average confidence interval of 37 cm, corresponding to roughly 12 Mbp in the chicken genome.

Table 3 Genetic mapping studies of chicken bone traits, including what type of study design (linkage or genome-wide association), animals, and traits were used.

Study	Type of study	Chickens	Traits
Schreiweis et al. (2005)	Linkage mapping	F_2 cross of broiler and White Leghorn	Tibial, fibular and humeral bone mineral morphology, bone mineral density and content (dual X-ray absorptiometry) Tibial breaking strength
Dunn et al. (2007)	Linkage mapping	F_2 cross of white layers divergently selected for bone index	Tibial breaking strength Humeral breaking strength Keel bone density (autoradiography)
Sharman et al. (2007)	Linkage mapping	F_2 cross of broiler and layer	Leg bowing Leg twisting Crooked digits Tibial morphology Tibial density (autoradiography) Tibial plateau angle Tibial torsion Tibial dyschondroplasia (autoradiography)
Rubin et al. (2007a), Wright et al. (2010)	Linkage mapping	F_2 cross of Red Junglefowl and White Leghorn	Femoral bone-mineral density and content of cortical and medullary bone (quantitative computed tomography) Femoral bone-mineral density and content (dual X-ray absorptiometry) Femoral bending and breaking strength Femoral torsional strength
Zhou et al. (2007)	Linkage mapping	F_2 cross of broiler and Leghorn F_2 cross of broiler and Fayoumi	Tibial bone-mineral density and content (dual X-ray absorptiometry)
Zhang et al. (2010, 2011)	Linkage mapping (chromosome 1)	F_2 cross of broiler and Baier	Femoral and keel length and weight
Podisi et al. (2012)	Linkage mapping	F_2 cross of broilers and layers	Humeral density (autoradiography)
Johnsson et al. (2014)	Linkage mapping	F_2 cross of Red Junglefowl and obsese layer line	Femoral bone-mineral density and content of cortical and medullary bone (quantitative computed tomography)
Johnsson et al. (2015)	Linkage mapping	Advanced intercross of Red Junglefowl and White Leghorn	Femoral bone-mineral density and content of cortical and medullary bone (quantitative computed tomography) Femoral gene expression (microarray)
Mignon-Grasteau et al. (2016)	Linkage mapping	F_2 cross of broilers divergently selected for digestive efficiency	Tibial morphology, dry weight, ash content Tibial breaking strength Excreted and serum phosphorous (colorimetry)
Guo et al. (2017)	GWAS	F_2 cross of While Leghorn and Dongxiang	Tibial breaking strength Femoral bone-mineral density and content (dual X-ray absorptiometry)
Raymond et al. (2018)	GWAS	White layers (Lohmann breeding programme)	Tibial breaking strength
Faveri et al. (2019)	Linkage mapping	F_2 cross of layer and broiler	Tibial morphology, dry-matter percentage, ash content, mineral composition (spectrophotometry)

5 Measuring bone traits

As seen in the literature review above, there are several different ways to measure bone traits. Skeletal defects except dyschondroplasia have been visually scored, while dyschondroplasia has been measured either with post-mortem scoring after dissecting the bone or by low-intensity X-ray imaging. The latter can also be done in the field (Bartels et al., 1989) on live birds (Eusemann et al., 2018). Bone-quality traits have been measured mostly by bone breaking, ashing (measuring the mineral fraction of bone and allowing spectroscopic analysis of the ash), autoradiography, dual X-ray absorptiometry and peripheral quantitative computed tomography. Alternatives that can be performed on live birds include keel bone palpation or ultrasound (Casey-Trott et al., 2015; Preisinger, 2018). It seems likely that future developments in imaging speed and cost will open new avenues in this area suitable for genetic selection.

Being able to measure the trait on the selection candidate themselves is a massive advantage for breeding. In a research context, post-mortem phenotyping may be acceptable (even for selection; Bishop et al. (2000), for example, used a retrospective selection method). Breeding on a trait that needs to be genotyped, post-mortem in a commercial setting would mean raising relatives for testing, and thus provides less accurate selection. For example, the femoral head-necrosis trait is more highly heritable than several other skeletal defects, but must be assessed post-mortem. Consequently, it still showed a flat genetic trend (Siegel et al., 2019), while less-heritable defects had decreasing genetic trends. However, other factors, such as unfavourable genetic correlations (which were not reported), may contribute to this difference.

Both morphological and X-ray methods give rise to high-dimensional data with many correlated features (Hocking et al., 2009; Wright et al., 2010), especially peripheral computed tomography, which allows for separate measurements on the cortical and trabecular/medullary fractions of bone. Measures from different bones add another correlated dimension. To be able to know the best bone phenotype to measure, we would need large quantitative genetic studies that compared them, and ideally also compared them between different production environments.

6 Summary

In summary, both skeletal defects and bone-quality traits in poultry are low to moderately heritable. Skeletal defects are only modestly unfavourably correlated with production traits, relatively easy to measure, and consequently show evidence of reduction in commercial breeding programmes. Bone-quality traits are more difficult to deal with, due largely to more involved phenotyping, and have therefore not seen similar improvements to those for defects. There are also potentially unfavourable correlations with egg traits, but this has been less well-explored in the research literature.

Linkage mapping and genome-wide association studies have identified numerous loci associated with bone traits. The genomic resolution is largely lacking to isolate individual causative genes and variants. However, several studies have identified plausible candidate genes located close to loci with strong association signals.

This current state of knowledge about poultry bone genetics suggests some future directions for research. There is a need for ways to measure bone strength and quality in live birds that can be deployed in breeding. There are also rapidly developing functional genomic methods that may help in identifying causative genes and variants for bone traits.

7 Future trends in research

There are at least three ways that future genetics research might benefit bone traits in chickens:

1. Identifying causative genes for bone quality can lead to biological insights that guide other interventions (dietary, management etc.).
2. Deep bone phenotyping (phenomics) and quantitative genetics can precisely quantify the relationship between different ways to measure bone, and identify good proxies for bone quality, that are relevant for management regimes used in production, and can be used in genomic selection for bone quality.
3. Functional genomics of bone tissues and cells can help identify molecular phenotypes (cellular phenotypes, metabolic markers, gene expression) and gene networks that underlie healthy bone in poultry.

7.1 Causative genes

Causative gene identification is hard, and even if we were able to identify causative variants for bone-related traits, it is not obvious how they should be used in breeding. Marker-assisted selection, while possible for major genes, is not likely to be feasible for complex traits. For the immediate future, improvements to skeletal traits are likely to come about through genomic selection (see other chapters in this volume).

However, causative gene identification could serve as an entry point in the biology of a trait, by highlighting biological functions to guide intervention. While some have argued that massive polygenicity and the complexity of gene regulation make causative genes unlikely to lead to easy biological insight (Boyle et al., 2017), genome-wide association for osteoporosis in humans have identified previously known drug targets (Richards et al., 2012). Similarly, causative gene identification in poultry may inspire new interventions for better bones.

The move from linkage mapping in crosses of divergent lines to genome-wide association within chicken breeding populations will improve genomic resolution for fine-mapping and make genetic mapping results more applicable to breeding populations. However, follow-up functional genomics and molecular biology will still be needed to identify causative genes. For example, it has taken 13 years from the publication of a locus (Dunn et al., 2007) to the elucidation of an eQTL at this locus and its physiological consequences (de Koning et al., 2020), although increasing sophistication of technology throughput and analysis should make this task easier in the future.

7.2 Bone phenomics

It is still an open question of how to best phenotype bone-quality traits in an efficient and informative way. If research could precisely quantify the genetic correlations between different options for measuring bone quality and their link with problems in different management regimes used in industry, this could make a major difference to poultry breeding and welfare. Furthermore, bone quality could be studied from a developmental perspective, and put in context with reproductive and behavioural traits. There are significant open challenges with both gathering and interpreting such high-dimensional data.

7.3 Functional genomics of avian bone

We are in the midst of an unprecedented explosion of genomic data. In addition to genome sequences of chicken populations, functional genomics data on chickens beyond the genome sequence level, including regulatory information such as open chromatin and chromatin immunoprecipitation sequencing, long read RNA-sequencing identifying splice isoforms and noncoding RNA, are becoming available. Some studies have measured transcriptome-wide gene expression in chicken bone (Horvat-Gordon et al., 2010; Johnsson et al., 2015; Rubin et al., 2007b). However, with the availability of high-throughput RNA sequencing, chromatin sequencing, even of single cells, much richer molecular data is potentially available.

There are examples from mice that use gene expression data to localise effects of putative causative genes to different bone cell types (Calabrese et al., 2017; Farber et al., 2011; Mesner et al., 2019), and examples from humans of fine-mapping genome-wide association signals by overlaying functional genomic data on chromatin accessibility and contacts (Morris et al., 2019).

In addition to aiding causative gene identification, there are classic studies, such as Clum et al.'s comparative study of chicken bone development (1995), that could be revisited with contemporary molecular methods. Coupled with a molecular understanding of which genes regulate skeletal patterning and bone-cell differentiation and proliferation, such studies could pinpoint the molecular processes underlying broiler and layer bone development.

8 Acknowledgement

I would like to thank Ian Dunn for helpful comments on the manuscript.

9 Where to look for further information

There are several reviews of skeletal problems in poultry (Bradshaw et al., 2002; Pines and Reshef, 2015; Toscano, 2018).

Furthermore, there is information about the development and molecular genetics of bone health in humans (Karsenty and Wagner, 2002; Richards et al., 2012; Rivadeneira and Mäkitie, 2016), which may be useful for putting results from chickens in perspective.

10 References

Bartels, J., McDaniel, G. and Hoerr, F. 1989. Radiographic diagnosis of tibial dyschondroplasia in broilers: a field selection technique. *Avian Diseases* 33(2), 254-7.

Bihan-Duval, E. L., Beaumont, C. and Colleau, J. 1997. Estimation of the genetic correlations between twisted legs and growth or conformation traits in broiler chickens. *Journal of Animal Breeding and Genetics* 114, 239-59.

Bishop, S., Fleming, R., McCormack, H., Flock, D. and Whitehead, C. 2000. Inheritance of bone characteristics affecting osteoporosis in laying hens. *British Poultry Science* 41, 33-40.

Bloom, M. A., Domm, L. V., Nalbandov, A. V. and Bloom, W. 1958. Medullary bone of laying chickens. *American Journal of Anatomy* 102, 411-53.

Boyle, E. A., Li, Y. I. and Pritchard, J. K. 2017. An expanded view of complex traits: from polygenic to omnigenic. *Cell* 169, 1177-86.

Bradshaw, R., Kirkden, R. and Broom, D. 2002. A review of the aetiology and pathology of leg weakness in broilers in relation to welfare. *Avian and Poultry Biology Reviews* 13, 45-103.

Calabrese, G. M., Mesner, L. D., Stains, J. P., Tommasini, S. M., Horowitz, M. C., Rosen, C. J. and Farber, C. R. 2017. Integrating GWAS and co-expression network data identifies bone mineral density genes SPTBN1 and MARK3 and an osteoblast functional module. *Cell Systems* 4, 46-59.

Casey-Trott, T., Heerkens, J., Petrik, M., Regmi, P., Schrader, L., Toscano, M. J. and Widowski, T. 2015. Methods for assessment of keel bone damage in poultry. *Poultry Science* 94, 2339-50.

Clum, N., McClearn, D. and Barbato, G. 1995. Comparative embryonic development in chickens with different patterns of postnatal growth. *Growth, Development, and Aging: GDA* 59, 129-38.

Cransberg, P., Parkinson, G., Wilson, S. and Thorp, B. 2001. Sequential studies of skeletal calcium reserves and structural bone volume in a commercial layer flock. *British Poultry Science* 42, 260–5.

Dacke, C., Arkle, S., Cook, D., Wormstone, I., Jones, S., Zaidi, M. and Bascal, Z. 1993. Medullary bone and avian calcium regulation. *Journal of Experimental Biology* 184, 63–88.

de Koning, D.-J., Dominguez-Gasca, N., Fleming, R., Gill, A., Kurian, D., Law, A., McCormack, H., Morrice, D., Sanchez-Rodriguez, E., Rodriguez-Navarro, A., Preisinger, R., Schmutz, M., Smídová, V., Turner, F., Wilson, P., Zhou, R. and Dunn, I. 2020. An eQTL in the cystathionine beta synthase gene is linked to osteoporosis in laying hens. *Genetics Selection Evolution* 52(1), 13.

Dunn, I., Fleming, R., McCormack, H., Morrice, D., Burt, D., Preisinger, R. and Whitehead, C. 2007. A QTL for osteoporosis detected in an F2 population derived from White Leghorn chicken lines divergently selected for bone index. *Animal Genetics* 38, 45–9.

Eusemann, B. K., Baulain, U., Schrader, L., Thöne-Reineke, C., Patt, A. and Petow, S. 2018. Radiographic examination of keel bone damage in living laying hens of different strains kept in two housing systems. *PLoS ONE* 13.

Farber, C. R., Bennett, B. J., Orozco, L., Zou, W., Lira, A., Kostem, E., Kang, H. M., Furlotte, N., Berberyan, A. and Ghazalpour, A. 2011. Mouse genome-wide association and systems genetics identify Asxl2 as a regulator of bone mineral density and osteoclastogenesis. *PLoS Genetics* 7, e1002038.

Farquharson, C. and Jefferies, D. 2000. Chondrocytes and longitudinal bone growth: the development of tibial dyschondroplasia. *Poultry Science* 79, 994–1004.

Faveri, J., Pinto, L., de Camargo, G., Pedrosa, V., Peixoto, J., Marchesi, J., Kawski, V., Coutinho, L. and Ledur, M. 2019. Quantitative trait loci for morphometric and mineral composition traits of the tibia bone in a broiler× layer cross. *Animal* 13(8), 1563–9.

Fleming, R., McCormack, H., McTeir, L. and Whitehead, C. 2004. Incidence, pathology and prevention of keel bone deformities in the laying hen. *British Poultry Science* 45, 320–30.

Fleming, R., McCormack, H., McTeir, L. and Whitehead, C. 2006. Relationships between genetic, environmental and nutritional factors influencing osteoporosis in laying hens. *British Poultry Science* 47, 742–55.

González-Cerón, F., Rekaya, R. and Aggrey, S. 2015a. Genetic analysis of bone quality traits and growth in a random mating broiler population. *Poultry Science* 94, 883–9.

González-Cerón, F., Rekaya, R. and Aggrey, S. 2015b. Genetic relationship between leg problems and bone quality traits in a random mating broiler population. *Poultry Science* 94, 1787–90.

Guo, J., Sun, C., Qu, L., Shen, M., Dou, T., Ma, M., Wang, K. and Yang, N. 2017. Genetic architecture of bone quality variation in layer chickens revealed by a genome-wide association study. *Scientific Reports* 7, 45317.

Hocking, P., Bain, M., Channing, C., Fleming, R. and Wilson, S. 2003. Genetic variation for egg production, egg quality and bone strength in selected and traditional breeds of laying fowl. *British Poultry Science* 44, 365–73.

Hocking, P., Sandercock, D., Wilson, S. and Fleming, R. 2009. Quantifying genetic (co) variation and effects of genetic selection on tibial bone morphology and quality in 37 lines of broiler, layer and traditional chickens. *British Poultry Science* 50, 443–50.

Horvat-Gordon, M., Praul, C., Ramachandran, R., Bartell, P. and Leach Jr., R. 2010. Use of microarray analysis to study gene expression in the avian epiphyseal growth plate. *Comparative Biochemistry and Physiology Part D: Genomics and Proteomics* 5, 12-23.
Johnsson, M., Rubin, C., Höglund, A., Sahlqvist, A., Jonsson, K. B., Kerje, S., Ekwall, O., Kämpe, O., Andersson, L. and Jensen, P. 2014. The role of pleiotropy and linkage in genes affecting a sexual ornament and bone allocation in the chicken. *Molecular Ecology* 23, 2275-86.
Johnsson, M., Jonsson, K. B., Andersson, L., Jensen, P. and Wright, D. 2015. Genetic regulation of bone metabolism in the chicken: similarities and differences to mammalian systems. *PLoS Genetics* 11, e1005250.
Johnsson, M., Jonsson, K. B., Andersson, L., Jensen, P. and Wright, D. 2016. Quantitative trait locus and genetical genomics analysis identifies putatively causal genes for fecundity and brooding in the chicken. *G3: Genes, Genomes, Genetics* 6, 311-19.
Kapell, D., Hill, W., Neeteson, A.-M., McAdam, J., Koerhuis, A. and Avendano, S. 2012. Twenty-five years of selection for improved leg health in purebred broiler lines and underlying genetic parameters. *Poultry Science* 91, 3032-43.
Karsenty, G. and Wagner, E. F. 2002. Reaching a genetic and molecular understanding of skeletal development. *Developmental Cell* 2, 389-406.
Kestin, S., Su, G. and Sorensen, P. 1999. Different commercial broiler crosses have different susceptibilities to leg weakness. *Poultry Science* 78, 1085-90.
Knott, L., Whitehead, C. C., Fleming, R. and Bailey, A. 1995. Biochemical changes in the collagenous matrix of osteoporotic avian bone. *Biochemical Journal* 310, 1045-51.
Kuhlers, D. and McDaniel, G. 1996. Estimates of heritabilities and genetic correlations between tibial dyschondroplasia expression and body weight at two ages in broilers. *Poultry Science* 75, 959-61.
Le Bihan-Duval, E., Beaumont, C. and Colleau, J. 1996. Genetic parameters of the twisted legs syndrome in broiler chickens. *Genetics Selection Evolution* 28, 177.
Mercer, J. and Hill, W. 1984. Estimation of genetic parameters for skeletal defects in broiler chickens. *Heredity* 53, 193.
Mesner, L. D., Calabrese, G. M., Al-Barghouthi, B., Gatti, D. M., Sundberg, J. P., Churchill, G. A., Godfrey, D. A., Ackert-Bicknell, C. L. and Farber, C. R. 2019. Mouse genome-wide association and systems genetics identifies Lhfp as a regulator of bone mass. *PLoS Genetics* 15, e1008123.
Mignon-Grasteau, S., Chantry-Darmon, C., Boscher, M.-Y., Sellier, N., Chabault-Dhuit, M., Le Bihan-Duval, E. and Narcy, A. 2016. Genetic determinism of bone and mineral metabolism in meat-type chickens: a QTL mapping study. *Bone Reports* 5, 43-50.
Morris, J. A., Kemp, J. P., Youlten, S. E., Laurent, L., Logan, J. G., Chai, R. C., Vulpescu, N. A., Forgetta, V., Kleinman, A. and Mohanty, S. T. 2019. An atlas of genetic influences on osteoporosis in humans and mice. *Nature Genetics* 51, 258.
Pines, M. and Reshef, R. 2015. Poultry bone development and bone disorders. In: Scanes, C.G. (Ed.), *Sturkie's Avian Physiology*. Elsevier, pp. 367-77.
Podisi, B., Knott, S., Dunn, I., Burt, D. and Hocking, P. 2012. Bone mineral density QTL at sexual maturity and end of lay. *British Poultry Science* 53, 763-9.
Preisinger, R. 2018. Innovative layer genetics to handle global challenges in egg production. *British Poultry Science* 59, 1-6.
Raymond, B., Johansson, A. M., McCormack, H. A., Fleming, R. H., Schmutz, M., Dunn, I. C. and De Koning, D. J. 2018. Genome-wide association study for bone strength in laying hens. *Journal of Animal Science* 96, 2525-35.

Rekaya, R., Sapp, R., Wing, T. and Aggrey, S. 2013. Genetic evaluation for growth, body composition, feed efficiency, and leg soundness. *Poultry Science* 92, 923-9.

Riber, A. B., Casey-Trott, T. M. and Herskin, M. S. 2018. The influence of keel bone damage on welfare of laying hens. *Frontiers in Veterinary Science* 5, 6.

Richards, J. B., Zheng, H.-F. and Spector, T. D. 2012. Genetics of osteoporosis from genome-wide association studies: advances and challenges. *Nature Reviews Genetics* 13, 576.

Riddell, C. 1976. Selection of broiler chickens for a high and low incidence of tibial dyschondroplasia with observations on spondylolisthesis and twisted legs (perosis). *Poultry Science* 55, 145-51.

Rivadeneira, F. and Mäkitie, O. 2016. Osteoporosis and bone mass disorders: from gene pathways to treatments. *Trends in Endocrinology and Metabolism* 27, 262-81.

Rodriguez-Navarro, A., McCormack, H., Fleming, R., Alvarez-Lloret, P., Romero-Pastor, J., Dominguez-Gasca, N., Prozorov, T. and Dunn, I. 2018. Influence of physical activity on tibial bone material properties in laying hens. *Journal of Structural Biology* 201, 36-45.

Rubin, C., Brändström, H., Wright, D., Kerje, S., Gunnarsson, U., Schutz, K., Fredriksson, R., Jensen, P., Andersson, L. and Ohlsson, C. 2007a. Quantitative trait loci for BMD and bone strength in an intercross between domestic and wildtype chickens. *Journal of Bone and Mineral Research* 22, 375-84.

Rubin, C.-J., Lindberg, J., Fitzsimmons, C., Savolainen, P., Jensen, P., Lundeberg, J., Andersson, L. and Kindmark, A. 2007b. Differential gene expression in femoral bone from red junglefowl and domestic chicken, differing for bone phenotypic traits. *BMC Genomics* 8, 208.

Schreiweis, M. A., Hester, P. Y. and Moody, D. E. 2005. Identification of quantitative trait loci associated with bone traits and body weight in an F2 resource population of chickens. *Genetics Selection Evolution* 37, 677.

Sharman, P., Morrice, D., Law, A., Burt, D. and Hocking, P. 2007. Quantitative trait loci for bone traits segregating independently of those for growth in an F2 broiler× layer cross. *Cytogenetic and Genome Research* 117, 296-304.

Siegel, P., Barger, K. and Siewerdt, F. 2019. Limb health in broiler breeding: history using genetics to improve welfare. *The Journal of Applied Poultry Research* 28(4), 785-90.

Sparke, A., Sims, T., Avery, N., Bailey, A., Fleming, R. and Whitehead, C. 2002. Differences in composition of avian bone collagen following genetic selection for resistance to osteoporosis. *British Poultry Science* 43, 127-34.

Stratmann, A., Fröhlich, E., Gebhardt-Henrich, S., Harlander-Matauschek, A., Würbel, H. and Toscano, M. J. 2016. Genetic selection to increase bone strength affects prevalence of keel bone damage and egg parameters in commercially housed laying hens. *Poultry Science* 95, 975-84.

Toscano, M. 2018. Skeletal problems in contemporary commercial laying hens. In: Mench, J. (Ed.), *Advances in Poultry Welfare*. Elsevier, pp. 151-73.

Van de Velde, J., Vermeiden, J., Touw, J. and Veldhuijzen, J. 1984. Changes in activity of chicken medullary bone cell populations in relation to the egg-laying cycle. *Metabolic Bone Disease and Related Research* 5, 191-3.

Van Goor, A., Ashwell, C. M., Persia, M. E., Rothschild, M. F., Schmidt, C. J. and Lamont, S. J. 2016. Quantitative trait loci identified for blood chemistry components of an advanced intercross line of chickens under heat stress. *BMC Genomics* 17, 287.

Whitehead, C. 2004. Overview of bone biology in the egg-laying hen. *Poultry Science* 83, 193-9.

Wilson, S., Duff, S. and Whitehead, C. 1992. Effects of age, sex and housing on the trabecular bone of laying strain domestic fowl. *Research in Veterinary Science* 53, 52-8.

Wong-Valle, J., McDaniel, G., Kuhlers, D. and Bartels, J. 1993. Correlated responses to selection for high or low incidence of tibial dyschondroplasia in broilers. *Poultry Science* 72, 1621-9.

Wright, D., Rubin, C., Barrio, A. M., Schütz, K., Kerje, S., Brändström, H., Kindmark, A., Jensen, P. and Andersson, L. 2010. The genetic architecture of domestication in the chicken: effects of pleiotropy and linkage. *Molecular Ecology* 19, 5140-56.

Wright, D., Rubin, C., Schutz, K., Kerje, S., Kindmark, A., Brandström, H., Andersson, L., Pizzari, T. and Jensen, P. 2012. Onset of sexual maturity in female chickens is genetically linked to loci associated with fecundity and a sexual ornament. *Reproduction in Domestic Animals* 47, 31-6.

Yalcin, S., Zhang, X., McDaniel, G. and Kuhlers, D. 2000. Effects of divergent selection for incidence of tibial dyschondroplasia (TD) on purebred and crossbred performance. 2. Processing yield. *British Poultry Science* 41, 566-9.

Zhang, X., McDaniel, G., Yalcin, Z. and Kuhlers, D. 1995. Genetic correlations of tibial dyschondroplasia incidence with carcass traits in broilers. *Poultry Science* 74, 910-15.

Zhang, X., McDaniel, G., Roland, D. and Kuhlers, D. 1998. Response to ten generations of divergent selection for tibial dyschondroplasia in broiler chickens: growth, egg production, and hatchability. *Poultry Science* 77, 1065-72.

Zhang, H., Zhang, Y., Wang, S., Liu, X., Zhang, Q., Tang, Z. and Li, H. 2010. Detection and fine mapping of quantitative trait loci for bone traits on chicken chromosome one. *Journal of Animal Breeding and Genetics* 127, 462-8.

Zhang, H., Liu, S., Zhang, Q., Zhang, Y., Wang, S., Wang, Q., Wang, Y., Tang, Z. and Li, H. 2011. Fine-mapping of quantitative trait loci for body weight and bone traits and positional cloning of the *RB1* gene in chicken. *Journal of Animal Breeding and Genetics* 128, 366-75.

Zhou, H., Deeb, N., Evock-Clover, C., Mitchell, A., Ashwell, C. and Lamont, S. J. 2007. Genome-wide linkage analysis to identify chromosomal regions affecting phenotypic traits in the chicken. III. Skeletal integrity. *Poultry Science* 86, 255-66.

Chapter 2

Leg disorders in poultry: bacterial chondronecrosis with osteomyelitis (BCO)

Robert F. Wideman Jr., University of Arkansas, USA

1 Introduction

Selection for rapid growth has consistently been implicated in the susceptibility of broilers and turkeys to leg weakness (Wise, 1970a; Nestor, 1984; Kestin et al., 1999, 2001; Sorensen, 1992; Nestor and Anderson, 1998). The highest incidences of leg weakness typically occur in flocks with the fastest growth, and lameness incidences usually are reduced by management strategies that decelerate early growth rates (Riddell, 1983; Havenstein et al., 1994; McNamee et al., 1999; Williams et al., 2000; Kestin et al., 2001; Bradshaw et al., 2002; Julian, 2005).

Growth-associated leg weakness encompasses a variety of distinct pathologies including

- tibial dyschondroplasia (Leach and Nesheim, 1965; Riddell, 1976; Sheridan et al., 1978; Walser et al., 1982; Zhang et al., 1995; Kuhlers and McDaniel, 1996),
- spondylolisthesis (Wise, 1970b, 1973; Riddell, 1973, 1976; Khan et al., 1977),
- valgus-varus deformities (Mercer and Hill, 1984; Sorensen, 1992),

http://dx.doi.org/10.19103/AS.2016.0011.23
Published by Burleigh Dodds Science Publishing Limited, 2017

- twisted leg or perosis (Somes, 1969; Haye and Simons, 1978; Mercer and Hill, 1984; Sorensen, 1992),
- bacterial osteomyelitis (McNamee et al., 1998; McNamee and Smyth, 2000; Wideman et al., 2012, 2013).

These distinctively different pathologies are linked by the hypothesis that, in susceptible subsets of the population, the skeleton does not mature rapidly enough to support exponential increases in body mass (Wise, 1970a; Kestin et al., 2001). The concept of 'skeletal maturation' broadly includes processes such as chondrocyte proliferation and differentiation within the growth plates, bone mineralization and remodelling, vasculogenesis and angiogenesis, and tendon and ligament development.

This chapter focuses on a specific category of lameness known as bacterial chondronecrosis with osteomyelitis (BCO), widely recognized as one of the most common causes of lameness in broilers (McNamee and Smyth, 2000). The pathogenesis of BCO reflects the fact that bacterial infection and necrosis typically develop in bones that are subjected to repeated mechanical stress, primarily but not exclusively, including the proximal femora, the proximal tibiae and the flexible thoracic vertebrae (Wise, 1971; Nairn, 1973; Julian, 1985; Wyers et al., 1991; McNamee et al., 1998; McNamee and Smyth, 2000; Stalker et al., 2010; Kense and Landman, 2011; Wideman et al., 2012; Wideman and Prisby, 2013). Several topics are addressed. The pathogenesis of BCO is summarized. Experimental models that successfully trigger BCO are reviewed. Sources of bacteria that infect the bones are discussed, with the primary emphasis on bacterial translocation across the gastrointestinal epithelium. Finally, we review the efficacy of probiotics as a prophylactic treatment for BCO.

2 The pathogenesis of BCO

There are a number of stages which lead to BCO. High growth rates in poultry lead to the development of unstable growth plates in bone. When subjected to mechanical stress, these develop microfractures which allow bacteria to infect the growth plate. The resulting damage leads to lameness.

2.1 *Osteochondrosis*

Growth of the leg bones includes elongation accomplished via growth plates located at both ends of the shaft (diaphysis), as well as marked increases in the overall diameter attributable to highly dynamic remodelling of cortical bone (Wideman and Prisby, 2013). As demonstrated by Applegate and Lilburn (2002), the femora and tibiae increase approximately fourfold in length during the first six weeks posthatch, with mid-shaft diameters increasing three- to fivefold during the same interval. Other investigators have published similar estimates of rapid leg bone growth in broilers (Wise, 1970a; Riddell, 1975; Thorp, 1988c; Bond et al., 1991; Williams et al., 2000; Yalcin et al., 2001; Yair et al., 2012). As shown in Fig. 1E, the growing ends of avian bones are comprised of a cartilagenous epiphysis, a growth plate or physis consisting of columns of chondrocytes (cartilage cells) in sequential maturational layers and a metaphysis consisting of calcifying chondrocytes and the newly formed spicules of trabecular bone that provide a support scaffolding for the growth plate. The pace of bone elongation is determined by the rate of mitosis within the proliferating zone of the growth plate, which, as it thickens, expands the physeal-epiphyseal boundary longitudinally at the end of the shaft (Wideman and Prisby, 2013). Maximal growth rates in poultry are associated with the presence of unusually long, unevenly aligned and mechanically fragile columns of unmineralized chondrocytes extending from the growth plate into the metaphysis. The proximal ends of the leg bones elongate twice as fast and have significantly thicker and more fragile growth plates when compared with the distal ends of the same bones (Hales, 1727; Church and Johnson, 1964; Lutfi, 1970a; Kember and Kirkwood, 1987; Leach and Gay, 1987; Thorp, 1988c; Kirkwood et al., 1989a,b; Kember et al., 1990).

Mechanical stress chronically exerted on thick, unstable growth plates can create osteochondrotic clefts or microfractures between and within the cartilage layers (e.g. epiphyseal-physeal osteochondrosis or osteochondrosis dissecans). The microfractures often transect local blood vessels, thereby causing focal ischaemia and necrosis. Osteochondrosis is routinely observed during histological evaluations of the leg bones and vertebrae of apparently healthy broilers and turkeys, leading to the assumption that osteochondrotic microfractures of the epiphyseal-physeal cartilage are a common noninfective, noninflammatory skeletal disorder (Wise, 1970b, 1973; McCaskey et al., 1982; Riddell et al., 1983; Duff, 1990a,b; Thorp, 1994). However, osteochondrotic clefts and associated regions of focal necrosis clearly constitute wound sites with an exposed collagen matrix that facilitates colonization by opportunistic bacterial pathogens. Accordingly, although clinical lameness has not been directly correlated with osteochondrosis per se, the resulting wound sites in fragile growth plates do provide suitable niches for secondary bacterial infections, thereby enhancing the susceptibility to osteomyelitis (Carnaghan, 1966; Wise, 1971; McNamee et al., 1998; Martin et al., 2011; Wideman and Prisby, 2013; Wideman, 2014).

2.2 Haematogenous bacterial distribution

Bacteria are transmitted to chicks from breeder parents, contaminated egg shells or hatchery sources (Skeeles, 1997; McCullagh et al., 1998; Rodgers et al., 1999; McNamee and Smyth, 2000; Stalker et al. 2010; Kense and Landman, 2011) or enter the chick's circulation via translocation through the integument, respiratory system or gastrointestinal tract (Mutalib et al., 1983a,b; Andreasen et al., 1993; Thorp et al., 1993; McNamee et al., 1999; Martin et al., 2011). They spread haematogenously and can reach the growth plate via terminal capillary loops that have a discontinuous (fenestrated) endothelium with openings large enough to permit cellular elements in the blood to pass into the cartilaginous matrix (Beaumont, 1967; Lutfi, 1970b; Hunt et al., 1979; Howlett, 1980; Howlett et al., 1984; Emslie and Nade, 1983, 1985). Haematogenously distributed bacteria possessing the specific ability to bind to exposed bone collagen appear in some cases to be more virulent in their capacity to trigger osteomyelitis (Kibenge et al., 1983; Smeltzer and Gillaspy, 2000). The bacteria form obstructive emboli in the epiphyseal and metaphyseal vascular plexuses, adhere directly to the exposed cartilage matrix and colonize osteochondrotic clefts and zones of necrosis. Bacterial foci and sequestrae within infected bone tissue are notoriously inaccessible to antibiotics and the immune system (Emslie and Nade, 1983; Emslie et al., 1983, 1984; Kibenge et al., 1983; Speers and Nade, 1985; Alderson et al., 1986a,b; Alderson and Nade, 1987; Thorp, 1988b; Thorp et al., 1993; McNamee et al., 1998, 1999; McNamee and Smyth, 2000; Smeltzer and Gillaspy, 2000; Kense and Landman, 2011; Martin et al., 2011).

2.3 Bacterial osteomyelitis and clinical lameness

Bacterial sequestrae rapidly expand into focal zones of necrosis or large fibrinonecrotic abscesses in the metaphysis of infected bones, destroying the vasculature and eliminating struts of metaphyseal trabecular bone that normally provide structural support to the growth plates (Emslie et al., 1983, 1984; Emslie and Nade, 1983, 1985; Alderson et al., 1986b; Thorp, 1988a; Daum et al., 1990; Wyers et al., 1991; Thorp et al., 1993; Skeeles, 1997; McNamee et al., 1998; Joiner et al., 2005; Wideman et al., 2012; Wideman and Prisby, 2013). Terminal BCO presents as necrotic degeneration and bacterial infection primarily within the proximal ends of the femora and tibiae, as well as in the growth plates of the flexible thoracic vertebrae (Fig. 1 and 2).

High incidences of femoral, tibial and vertebral BCO lesions have been observed in lame broilers from commercial flocks. The distal ends of the femora and tibiae are affected less frequently (Emslie et al., 1983; Thorp and Waddington, 1997; McNamee et al., 1999). It is not unusual for BCO to affect only one leg while the contralateral leg appears macroscopically normal

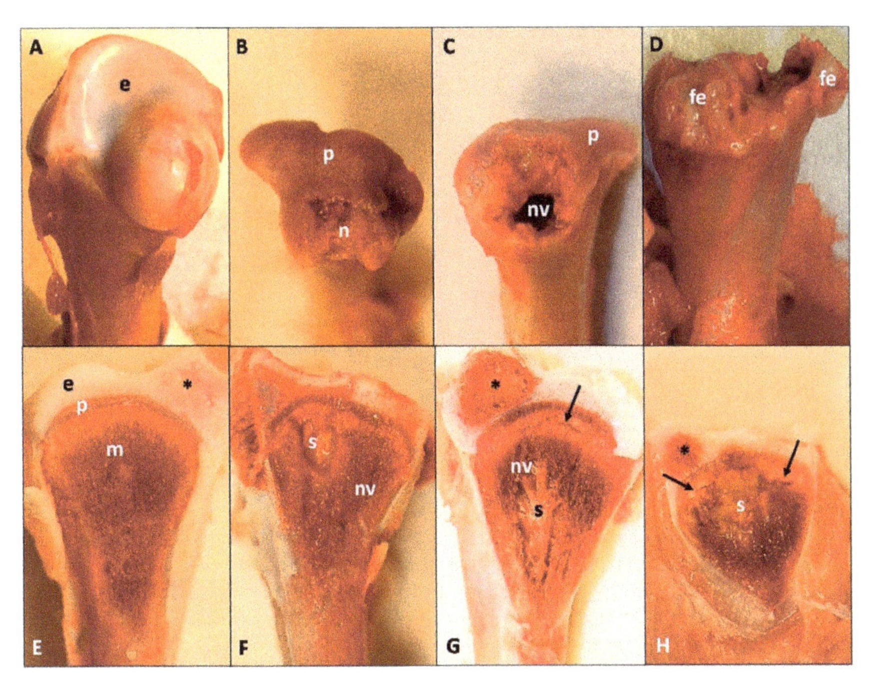

Figure 1 Stages of proximal femoral degeneration (**A–D**) and proximal tibial degeneration (**E–H**) leading progressively to BCO. (**A)** Normal proximal femur with white cap of epiphyseal cartilage (e). (**B**) Femoral head separation (FHS: epiphyseolysis) with the epiphysis remaining in the socket when the femur was disarticulated, revealing the underlying surface of the growth plate or physis (p) and an early region of necrosis (n). (**C**) Fracturing of the growth plate (p) revealing a necrotic void (nv) within the metaphysis. (**D**) Terminal femoral head necrosis in which the femoral epiphysis, physis and most of the metaphysis remained attached to the acetabulum when the diaphysis weakened by widely dispersed necrosis was fractured during disarticulation, revealing copious fibrinonecrotic exudate (fe). (**E**) Normal proximal tibia showing the epiphysis (e) with a secondary centre of ossification (*) and the physis/growth plate (p) fully supported by struts of trabecular bone in the metaphyseal zone (m). (**F–H**) Bacterial infiltration and sequestrae (s), necrotic voids (nv) and microfractures below the growth plate (arrows) provide macroscopic evidence of bone damage associated with osteomyelitis.

(McNamee et al., 1998; Dinev, 2009; Wideman, 2014). Femoral, tibial and vertebral BCO lesions usually occur together in commercial flocks, although in some outbreaks femoral and tibial lesions may be more common than vertebral lesions, whereas the opposite may be true in other outbreaks (Stalker et al., 2010; Kense and Landman, 2011; Wideman, personal observations).

Multiple opportunistic organisms have been isolated from BCO lesions, including several *Staphylococcus* spp., *Escherichia coli* and *Enterococcus cecorum*, often in mixed cultures with other bacteria, including *Salmonella* spp. (Smith, 1954; Carnaghan, 1966; Wise, 1971; Nairn and Watson, 1972;

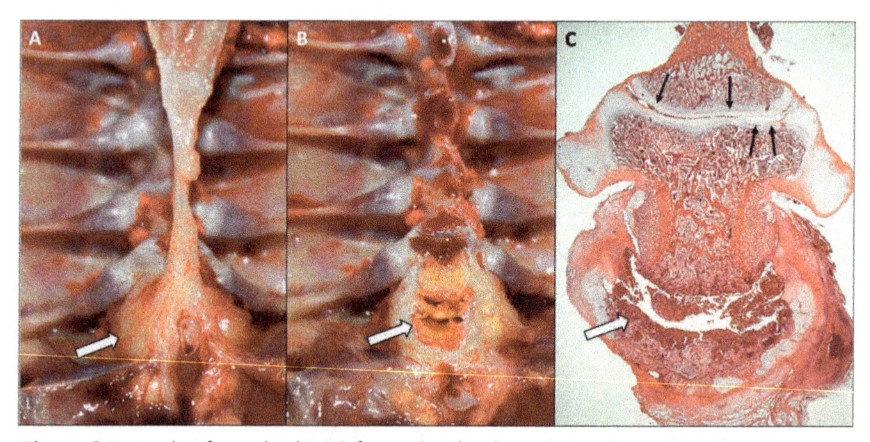

Figure 2 Example of vertebral BCO from a broiler that exhibited spinal cord compression and paraplegic hock-resting posture due to grossly abscessed vertebral bodies (white arrows). (**A**) Externally the flexible thoracic vertebral body exhibited marked nodular swelling with a distinctive yellow discolouration of the translucent remnants of the vertebral bodies. (**B**) Internally the caseous abscessation filled the caudal portion of the vertebral body. (**C**) Histologically the caudal epiphysis and physis are absent and have been replaced by the necrotic abscess; areas of osteochondrosis in the cranial articular cartilage layers are indicated by black arrows.

Griffiths et al., 1984; Andreasen et al., 1993; Tate et al., 1993; Thorp et al., 1993; McNamee et al., 1998; Butterworth, 1999; Smeltzer and Gillaspy, 2000; Joiner et al., 2005; Dinev, 2009; Stalker et al., 2010; Kense and Landman, 2011; Martin et al., 2011; Wijetunge et al., 2012).

Recently, Jiang et al. (2015) used molecular profiling of 16S ribosomal RNA (16S rRNA) gene sequences to comprehensively analyse the structure and diversity of microbial communities in the proximal femora and tibiae from clinically healthy broilers and from lame broilers with obvious BCO lesions. Complex microbial communities were detected in all samples, including bones that appeared to be macroscopically normal. Further analyses demonstrated major differences in the microbial communities of different bones (femur versus tibia) and different lesion categories (macroscopically normal versus gross BCO lesions). The genus *Staphylococcus* was overrepresented in bones with BCO lesions, along with the genera *Enterobacter* and *Serrotia*.

These results indicate that complex microbial communities exist in the growth plates and metaphyses of macroscopically normal proximal femora and tibiae and that BCO lesions are associated with a less diverse community with overrepresentation by potentially pathogenic genera (Jiang et al., 2015). There is the likelihood that numerous bacterial species routinely enter the bloodstream and are sequestered in the growth plates, where they form a complex microbial community. An obvious source of these translocating bacteria is the intestinal microbiome (Jiang et al., 2015), although recent evidence also reveals the presence of microbial communities within the conducting airways and lower respiratory tract of clinically healthy poultry (Nehme et al., 2005; Shabbir et al., 2015; Sohail et al., 2015).

3 Understanding and treating BCO: the role of experimental models

The aetiology, pathogenesis and treatment strategies for BCO have been difficult to investigate because the incidence often is low or nonexistent in research flocks. Experimental models for triggering BCO are needed to confirm key triggering mechanisms and stressors, reveal innate limitations or susceptibilities and provide a reliable test bed for evaluating practical prophylactic and therapeutic strategies.

3.1 Pathogen exposure models

Attempts to initiate BCO via a respiratory challenge have been only marginally successful (Devriese et al., 1972; Nicoll and Jensen, 1986; Jensen et al., 1987; Mutalib et al., 1983b; McNamee et al., 1999; Martin et al., 2011), whereas BCO has repeatedly been reproduced by injecting broilers and turkeys intravenously with appropriate strains of *Staphylococcus* spp. in quantities sufficient to cause sustained bacteraemia. The artificially induced bacteraemia resembles naturally occurring sepsis. Terminal metaphyseal and epiphyseal blood vessels rapidly become occluded by bacterial emboli and phagocytic cells, leading to focal necrosis and bacterial colonization of the ischaemic cartilage. Intravenous pathogen injections typically trigger high incidences of lameness attributable to synovitis and BCO within 24–72 hours postinjection (Smith, 1954; Carnaghan, 1966; Wise, 1971; Nairn, 1973; Emslie and Nade, 1983, 1985; Emslie et al., 1983; Kibenge et al., 1983; Mutalib et al., 1983a; Griffiths et al., 1984; Alderson et al., 1986b; Daum et al., 1990; Martin et al., 2011).

Smith (1954) induced bacteraemia and lameness by injecting *Staphylococcus* spp. intravenously, but none of the strains produced any effect when injected intraperitoneally or subcutaneously. Wise (1971) isolated *Staphylococcus aureus* and *Staphylococcus albus* from vertebral BCO abscesses in a turkey and a broiler, respectively. When these isolates were inoculated intravenously into broilers, only the *S. albus* isolate triggered vertebral BCO lesions (Wise, 1971). Griffiths et al. (1984) isolated *S. aureus* from lame commercial broilers. When inoculated intravenously into healthy broilers, this isolate exhibited a broad tropism for rapidly growing bone and triggered vertebral BCO as well as proximal femoral and tibial BCO (Griffiths et al., 1984). Alderson and Nade (1987) re-created septic arthritis and focal osteomyelitis by injecting the hock joints of four-week-old broilers with a strain of *S. aureus* isolated from lame chickens. Al-Rubaye et al. (2014) isolated *Staphylococcus agnetis* from bone and blood samples of broilers with clinical BCO. Administering this isolate in the drinking water increased the incidence of BCO threefold when compared with that in uninoculated broilers (Al-Rubaye et al., 2014).

During outbreaks of lameness in broiler and broiler breeder flocks in the European Union and North America, *E. cecorum* was implicated as the pathogen responsible for causing arthritis, tenosynovitis and femoral and vertebral BCO (Stalker et al., 2010; Kense and Landman, 2011; Martin et al., 2011; Wijetunge et al., 2012). Necropsies of lame broilers exhibiting a hock-sitting posture revealed nodular swelling, abscessation and osteomyelitis of the T4 vertebral body, accompanied by compression of the spinal cord. Lung tissue was focally adherent to the abscessed vertebrae. *E. coli* and Gram-positive cocci were isolated from the spinal abscesses, and the cocci were identified by 16S rRNA sequencing as *E. cecorum* (Stalker et al., 2010). Martin et al. (2011) used *E. cecorum* isolated from a commercial outbreak of vertebral BCO to inoculate male broiler breeder chicks via the air sac, and the intravenous and the oral routes. Oral transmission of *E. cecorum* was reported by these investigators to be the most productive method for reproducing the characteristic gross and microscopic lesions of vertebral BCO (Martin et al., 2011).

As a caveat to the use of direct pathogen exposure for triggering BCO, pre-existing osteochondrotic microfracturing of the epiphyseal-physeal cartilage may not be an essential permissive condition if naturally occurring sepsis or experimental pathogen inoculations cause bacteraemia. For example, intravenously injecting an effective dose of *Staphylococcus* spp. typically triggers sustained bacteraemia that causes bacterial emboli to form quite rapidly in the terminal epiphyseal and metaphyseal vascular plexuses throughout the skeleton, leading to widely dispersed osteomyelitis and a rapid (within one to three days) onset of lameness. This form of generalized/septic osteomyelitis apparently does not depend on the creation of osteochondrotic clefts (wound sites) as niches for bacterial colonization. Models of pathogen inoculation that very quickly overwhelm the body's defence mechanisms and trigger generalized septicaemia are unlikely to be useful for detecting differences in BCO susceptibility among broiler lines that may primarily depend on differences in innate susceptibility to osteochondrosis. Nor would such models be useful for demonstrating relative efficacies of practical prophylactic treatments that improve barrier integrity and thereby reduce bacterial translocation.

3.2 Mechanical models

These models for triggering BCO were designed to create sustained footing instability and thereby persistently exert excessive mechanical stress on susceptible leg and vertebral joints. Based on our current understanding, amplified footing instability creates conducive wound sites (e.g. osteochondrotic clefts, focal necrosis) in the epiphyseal-physeal cartilage. The mechanical approach avoids the need to purposefully inoculate the birds with known pathogens based on the assumption that opportunistic pathogens already are present but quiescent within the bird's tissues or in microbial communities or in the environment. For example, ramps placed under nipple waterers forced broilers to walk up a 30% slope to drink and, by eight weeks of age, triggered a higher cumulative incidence of BCO (27%) than the spontaneous (6%) incidence in pens without ramps (Wideman, 2014).

Portable obstacles known as 'speed bumps' have been used to consistently trigger BCO in broilers (Fig. 3) (Gilley et al., 2014). Speed bumps are constructed in the shape of an isosceles triangular prism, with sides having slopes ranging from 33 to 66%. They are designed to be installed between feeders and waterers in litter flooring facilities, thereby forcing the birds to climb up and then down the slopes as they move back and forth to eat and drink. Speed bumps trigger threefold higher incidences of BCO when compared with incidences in broilers reared on litter flooring without speed bumps (Gilley et al., 2014). Covering the entire surface of a pen floor with flat wire flooring panels subjects broilers to chronic footing instability as well as to behavioural stress attributable to denying the birds access to litter (*vide infra*). Incidences of lameness between 20 and 60% are reliably induced by this challenge, depending on the genetic susceptibility and hatchery source of the broilers being evaluated.

All of the mechanical models consistently triggered pathognomonic BCO lesion progression, primarily within the proximal femora and proximal tibiae. Vertebral BCO was rarely observed in these experiments until a recent study in which the chicks and feed were obtained from a commercial source. In that experiment, 80% of the lameness was attributed to proximal femoral and tibial BCO and 20% of the lameness was attributed to typical vertebral BCO lesions (Wideman et al., 2012, 2013, 2014, 2015a,b; Gilley et al., 2014; Prisby et al., 2014; Jiang et al., 2015; Wideman, personal observations).

Most of the lameness triggered by mechanical models develops between three and eight weeks of age, after broilers attain body weights sufficient to produce osteochondrotic microfractures, as has been reported for field outbreaks of BCO (McNamee et al., 1998; Dinev, 2009). Females tend to be marginally less susceptible to lameness than males when both genders are reared together (Wideman et al., 2014). Lameness progresses very rapidly in broilers that appeared to be healthy during the preceding 24–48 h, as previously reported by Joiner et al. (2005). Broilers tend to exhibit relatively mild BCO lesions when they are euthanized at the earliest onset of clinical symptoms (hesitancy to stand, eagerness to sit, slight wing-tip dipping), whereas birds that are permitted to live until they become fully immobilized (unable to move to feed or water) exhibit much more severe lesions.

Figure 3 Mechanical models for triggering BCO are designed to persistently exert excessive mechanical stresses on susceptible leg and vertebral joints, thereby creating conducive wound sites (e.g. osteochondrotic clefts, focal necrosis) in the epiphyseal-physeal cartilage. Portable obstacles known as 'speed bumps' (above) trigger threefold higher incidences of BCO when compared with those in broilers reared on litter flooring without speed bumps. Speed bumps are constructed in the shape of an isosceles triangular prism and are placed between feeders and waterers to force broilers to climb up and then down the slopes as they move back and forth to eat and drink (Gilley et al., 2014).

It is also apparent from necropsying survivors that severe lesions may occasionally be present in very large, apparently robust individuals that exhibit no signs of lameness or leg weakness (Wideman et al., 2012, 2014). We speculate that broilers and turkeys purposefully avoid exhibiting overt symptoms of lameness in order to avoid being victimized by the predatory behaviour of their flock mates. Indeed, gait scoring may not consistently or quantitatively reveal skeletal pathologies or abnormalities that are readily detected during the post-mortem examination of individual birds (Naas et al., 2009; Sandilands et al., 2011; Wideman, unpublished observations).

The pathogenesis of BCO cannot be instantaneous and apparently healthy broilers often possess subclinical lesions primarily consisting of the earliest macroscopic BCO lesion pathology. Subclinical lesions are equally likely to develop in males and females in left and right legs, and the status of the proximal femur does not determine the status of the ipsilateral or contralateral proximal tibia and vice versa (Wideman et al., 2012, 2013, 2014, 2015a,b; Gilley et al., 2014). These observations are consistent with the hypothesis that subclinical mechanical damage need not trigger overt lameness until the damaged area becomes infected. The resulting bacterial proliferation, immunological assault by responding phagocytes (macrophages and heterophils) and widespread lysis and necrosis of the metaphyseal trabecular bone and vasculature then culminate in intolerable discomfort and terminal lameness (Howlett, 1980; Duff, 1984b; Thorp et al., 1993; Wideman and Prisby, 2013).

Mechanical models have been used to generate practical strategies for reducing BCO incidences in commercial flocks. In two independent experiments, standard broiler crosses that grow rapidly at an early age developed higher incidences of lameness when compared with high yield crosses that initially tend to grow more slowly, directly demonstrating the association between rapid early growth and susceptibility to BCO (Wideman et al., 2013). Two independent experiments were also conducted to evaluate the influence of commercial sire lines on the susceptibility of broiler crosses to BCO. When grown to heavy yield weights, broilers derived from one sire line were consistently more likely to develop higher incidences of BCO than broilers derived from a second sire line, and these experimental results were confirmed by the experiences of broiler integrators using the same sires and crosses (Wideman et al., 2014).

The wire flooring model was used to demonstrate that enrofloxacin, a potent fluoroquinolone antimicrobial, is therapeutically effective for treating BCO (Wideman et al., 2015a). A proprietary 25-hydroxy vitamin D_3 product was also demonstrated to effectively reduce BCO incidences in broilers reared on wire flooring. This product is currently widely used prophylactically to minimize leg weakness in broilers and turkeys (Wideman et al., 2015b). By consistently triggering moderate to high incidences of BCO in research flocks, mechanical models have provided a reliable experimental protocol for comparing genetic lines and chick sources and for evaluating practical treatment strategies.

3.3 Stress-mediated immunosuppression models

Immunosuppression has been implicated in the aetiology of spontaneous BCO outbreaks in commercial poultry flocks (Mutalib et al., 1983a,b; Andreasen et al., 1993; Butterworth, 1999; McNamee et al., 1998, 1999; Huff et al., 2000;

McNamee and Smyth, 2000). Environmental stressors and immunosuppression contribute to the eruption of opportunistic pathogens in the proximal tibial joints of turkeys that develop turkey osteomyelitis complex (TOC). Bacterial infection of the proximal tibiae is a characteristic symptom of TOC, and in an experimental setting, TOC can be triggered by injecting turkey poults with repeated immunosuppressive doses of dexamethasone, a synthetic glucocorticoid (Wyers et al., 1991; Huff et al., 1998, 1999, 2000, 2005, 2006).

Based on the pathogenic similarities between BCO and TOC, we conducted three experiments to determine if dexamethasone injections might trigger BCO in broilers. In all three experiments, dexamethasone-injected broilers developed lameness primarily associated with lesions of the proximal tibiae (Wideman and Pevzner, 2012). Within this context it is highly likely that, in addition to creating footing instability, wire flooring models also elicit stress and immunosuppression. Elevated flooring systems that deprive birds of access to floor litter stimulate chronic stress, including immunosuppression (El-Lethey et al., 2003).

4 Sources and routes of bacterial colonization

4.1 Epidemiology

The current hypothesis for the pathogenesis of BCO proposes that exponential increases in early body weight impose excessive torque and shear stress on structurally immature epiphyseal and physeal cartilage, creating osteochondrotic wound sites that serve as ideal niches for colonization by haematogenously distributed opportunistic bacteria. This hypothesis does not specify the original sources of the bacterial pathogens. Sources and routes of entry of the offending bacterial pathogens may vary widely, as was suggested by a multiyear epidemiological study of sequential *Campylobacter jejuni* outbreaks in a broiler farm in the United Kingdom (Pearson et al., 1996). A contaminated water distribution system was identified as the initial source of a single *Campylobacter* serotype that was broadly distributed throughout all sheds on the farm. Marked overall reductions in *Campylobacter* isolation rates were achieved by cleaning and disinfecting the water lines and reservoirs, but thereafter, a limited number of different *Campylobacter* serotypes began to emerge intermittently. Notably these serotypes developed in some but not in all of the sheds. Specific *C. jejuni* serotypes were associated with the hatchery supplying the one-day-old chicks, as well as with grandparent flocks and broiler breeders. The investigators postulated that low-level vertical transmission from the broiler breeders to relatively few infected chicks would then lead to horizontal transmission of the same serotype throughout a shed (Pearson et al., 1996).

Outbreaks of BCO on so-called 'repeater' farms often exhibit similar patterns (Fig. 4). Incidences of femoral, tibial and vertebral BCO tend to vary widely among separate barns on the same farm, with some barns exhibiting little or no lameness and other barns develop high BCO incidences attributable to *Staphylococcus* spp. and *E. cecorum*. Different barns may be affected during sequential flock cycles, strongly implicating the breeder flock, hatchery or posthatch stress as potential factors contributing to early bacterial contamination. Flock managers often report that initially a few broilers develop BCO lameness at a focal location within a barn, after which lameness appears to spread horizontally throughout the affected barn but not into adjacent barns that are managed by the same personnel and provided feed from the same feed mill (Wideman, personal observations). Again, this epidemiological pattern appears to reflect low-level vertical transmission of pathogenic bacteria to a few chicks, followed by horizontal transmission of the same pathogen throughout a barn.

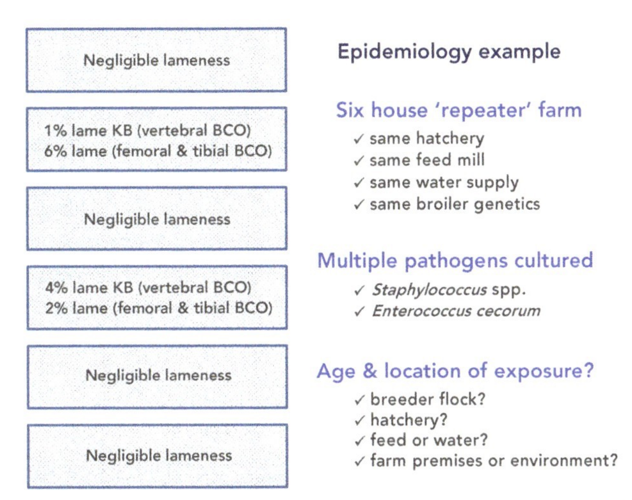

Figure 4 Outbreaks of BCO on so-called 'repeater' farms often exhibit similar patterns. Incidences of femoral, tibial and vertebral BCO tend to vary widely among separate barns on the same farm, with some barns exhibiting little or no lameness and other barns develop high BCO incidences attributable to *Staphylococcus* spp. and *E. cecorum*. Different barns may be affected during sequential flock cycles, strongly implicating the breeder flock or hatchery as potential factors contributing to early bacterial contamination. Flock managers often report that initially a few broilers develop BCO lameness at a focal location within a barn, after which lameness appears to spread horizontally throughout the affected barn but not into adjacent barns that are managed by the same personnel and provided feed from the same feed mill. This epidemiological pattern appears to reflect vertical transmission of pathogenic bacteria to a few chicks, followed by horizontal transmission of the same pathogen throughout a barn (Wideman, personal observations).

4.2 Vertical versus horizontal transmission

Breeder flocks and hatcheries have repeatedly been suspected of being the source of *Staphylococcus* spp. and *Enterococcus* spp. that subsequently are isolated from lame broilers (Skeeles, 1997; McCullagh et al., 1998; Rodgers et al., 1999; McNamee and Smyth, 2000; Stalker et al., 2010; Kense and Landman, 2011). Various bacterial species, including *Staphylococcus* and *Enterococcus*, have been isolated from spleens harvested from broiler embryos at day 18 of incubation. Broiler breeder hens are continuously subjected to stresses such as feed restriction, aggressive males, slat (nonlitter) flooring and oviposition. Sustained stress may promote bacterial translocation and bacteraemia sufficient to cause vertical transmission of the resident microbial communities from breeder hens to their progeny via the yolk follicles or reproductive tract (Humphrey and Lanning, 1988; Cox et al., 2002, 2005b; Liljebjelke et al., 2005; Funkhouser and Bordenstein, 2013). The broiler breeder hen's reproductive tract undergoes reverse peristalsis and therefore is susceptible to ascending infection by intestinal bacteria or contaminated semen, facilitating inoculation of the albumen and shell membranes before the shell is deposited (Reiber et al., 1995; Buhr et al., 2002; Cox et al., 2005a). Soiled eggshells can facilitate bacterial penetration through the shell and commensal colonization of the intestinal tracts of apparently healthy, newly hatched chicks (Clark and Bueschkens, 1985).

Opportunities for posthatch horizontal transmission abound. Certain hatcheries, particularly older hatcheries, are notorious for producing chicks that are susceptible to BCO. Our ongoing research using the wire flooring model repeatedly has implicated breeder flocks, hatchery sources or chick quality in the susceptibility of broilers to BCO. Indeed, our experimental chicks are routinely culled on day 14 because necropsies of runts and culls during the first two weeks posthatch often reveal evidence of systemic bacterial infection, including omphalitis, sepsis and osteomyelitis. Placing chicks on wire flooring on the day of hatch followed by heavy culling on day 14 has evolved into a standard experimental protocol for delaying the onset of clinical BCO until days 35–40 (Wideman et al., 2012, 2013, 2014, 2015a,b).

Current industry efforts to reduce the incidence of BCO in problematic (repeater) broiler complexes include improving hatchery sanitation, minimizing heat stress in the hatchery and during brooding, prescribing antibiotic treatments for chicks immediately posthatch and reducing the photoperiod to modestly reduce early growth performance. For these approaches to be effective, they must be addressing triggering mechanisms or early bacterial colonization that confer subclinical susceptibility at hatch or prior to hatching. For example, newly hatched chicks that had higher rectal temperatures or that were exposed to a high temperature in the hatcher on days 18 through 21 of incubation tended to develop higher incidences of BCO lameness on wire flooring when compared with chicks having normal rectal temperatures or that were exposed to a normal temperature in the hatcher (Fig. 5). These observations suggest that bacterial pathogens responsible for causing BCO may be present but quiescent very early pre- or posthatch, waiting for conducive wound sites to develop in larger heavier birds before clinical lameness attributable to BCO emerges. In that case, the epidemiology of BCO outbreaks beginning at 5 weeks of age would be difficult to understand if the primary source of their bacterial inoculation occurred *in ovo* or in the hatchery.

4.3 Bacterial translocation

For bacteria to be distributed haematogenously to osteochondrotic microfractures, they first must penetrate the body's defences and persist within the tissues. The ease with which viable bacteria can be routinely isolated from the blood and tissues of poultry pre- and posthatch readily discredits the common misconceptions that blood and tissues are essentially sterile in birds that appear to be clinically healthy and that the immune system efficiently destroys invading pathogens (Buhr et al., 2002; Cox et al., 2005a, 2006a,b, 2007; Clark et al., 2010; Richardson et al., 2011). In fact, broadly diverse microbial communities are now known to persist in the respiratory tract, unabsorbed yolk, and blood and bones of apparently healthy poultry, and these communities appear to be tolerated by the immune system (Nehme et al., 2005; Cox et al., 2006b,c; Jiang et al., 2015; Shabbir et al., 2015; Sohail et al., 2015).

Potential routes of bacterial penetration include vertical transmission from the breeder hen or horizontal transmission from the environment (e.g. the hatchery, growout facilities, accumulated litter, feed, water, bioaerosols) followed by translocation through the integument, respiratory epithelium or intestinal epithelium. Many poultry health professionals consider osteomyelitis to be a secondary infection that develops several weeks after a severe respiratory challenge (e.g. mycoplasma infection; poor hatchery sanitation; vigorous reaction to respiratory vaccine), an enteric challenge (e.g. coccidiosis; enteritis), an immunosuppressive challenge (e.g. stress; chicken anaemia virus; infectious bursal disease virus) or infection of the integument (e.g. wet navels; foot pad or hock lesions). Dermal lesions and omphalitis can lead to generalized sepsis and thus secondarily to broadly distributed, generalized osteomyelitis. For example, trimming the toes of recently hatched turkey poults provided a portal of entry for *S. aureus* to initiate bilateral osteomyelitis and synovitis in their femora and tibiae (Alfonso and Barnes, 2006). Also, injecting *S. aureus* directly into a broiler's hock joint triggered septic arthritis and bacterial colonization of the local epiphyseal and physeal vasculature (Alderson and Nade, 1987).

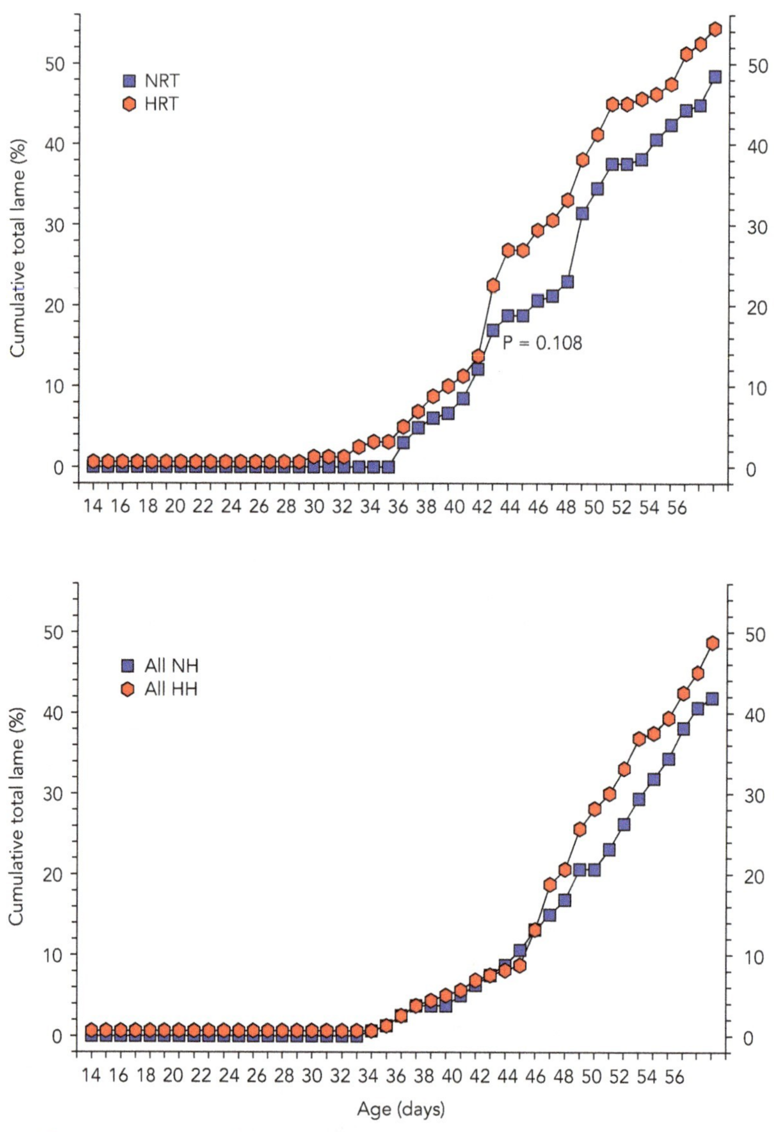

Figure 5 Newly hatched chicks were selected as having normal (<105.0 °F) or high (>105.5 °F) rectal temperatures (upper panel: NRT or HRT, respectively) in a commercial hatchery, or on days 18 through 21 of incubation the hatcher temperature in the University of Arkansas research hatchery was set to a normal (100 °F) or a high (101 °F) temperature (lower panel: NH or HH, respectively). The chicks were wing banded and co-mingled in pens with flat wire flooring from day 1 through day 56 to trigger high incidences of lameness attributable to BCO. Chicks with normal rectal temperatures or that were exposed to normal hatcher temperatures (NRT and NH) tended to have lower incidences of BCO lameness than chicks with high rectal temperatures or that were exposed to high hatcher temperatures (HRT and HH).

However, although compromised intestinal, respiratory and integument barriers may clearly precede and contribute to osteomyelitis outbreaks in commercial flocks, it is not necessary to induce mucosal barrier dysfunction as a precondition for triggering BCO in broilers with mechanical models or oral administration of pathogenic bacteria. Experiments have repeatedly demonstrated the ease with which orally administered or indigenous enteric bacteria can be translocated across the gastrointestinal epithelium to be carried by the circulatory system, perhaps adherent to or engulfed by phagocytes, to organs and tissues throughout the body (Lissner et al., 1983; Hand et al., 1984; Desmidt et al., 1998; Cossart and Sansonetti, 2004; Bailey et al., 2005; Cox et al., 2006b; Martin et al., 2011; Al-Rubaye et al., 2014).

5 The role of probiotics

Translocation of orally administered or indigenous enteric bacteria and haematogenous distribution by the circulatory system (likely engulfed by phagocytes) to organs and tissues throughout the body has been demonstrated repeatedly (Carter and Collins, 1974; Berg, 1992; Ando et al., 2000; Kremer et al., 2011; Richardson et al., 2011). Key components of the intestinal barrier to bacterial translocation are illustrated in Fig. 6 (upper panel). Goblet cells secrete protective mucus that lubricates the passage of digesta through the lumen and can hinder attachment by pathogenic bacteria to the glycocalyx of the brush border microvilli (Byrne et al., 2007). The inner mucin layer also serves as a matrix in which antimicrobial proteins (defensins) and secretory immunoglobulins (IgA) are sequestered to create an immune exclusion barrier protecting the luminal surfaces of the epithelium (Bar-Shira et al., 2014). Invasive bacteria that penetrate the mucus can bind to the brush border and induce phagocytosis by the enterocytes. Invasive bacteria may replicate intracellularly and can induce phagosomes to translocate transcellularly to the basal membrane of the cell, where viable bacteria emerge into the lamina propria to be phagocytized (but not neutralized) by macrophages (Cossart and Sansonetti, 2004; Byrne et al., 2007).

Most bacterial translocation across the intestinal barrier appears to utilize the paracellular route via 'leaky' tight junctions (Soriani et al., 2006). Tight junctions consisting of over 40 proteins (claudins, occludin, ZO-1) anchored to intracellular structural elements (including actin) comprise a key component of the intestinal barrier by sealing the apical perimeters of adjacent enterocytes (intestinal epithelial cells). New epithelial cells are continuously generated by stem cells in the crypts and differentiate into enterocytes or goblet cells during their 72–96 h migration from the crypts to the tip of a villus (Uni et al., 1995; Bohorquez et al., 2011). This continuous, life-long regeneration of the intestinal epithelium creates ongoing opportunities to acutely modulate the synthesis

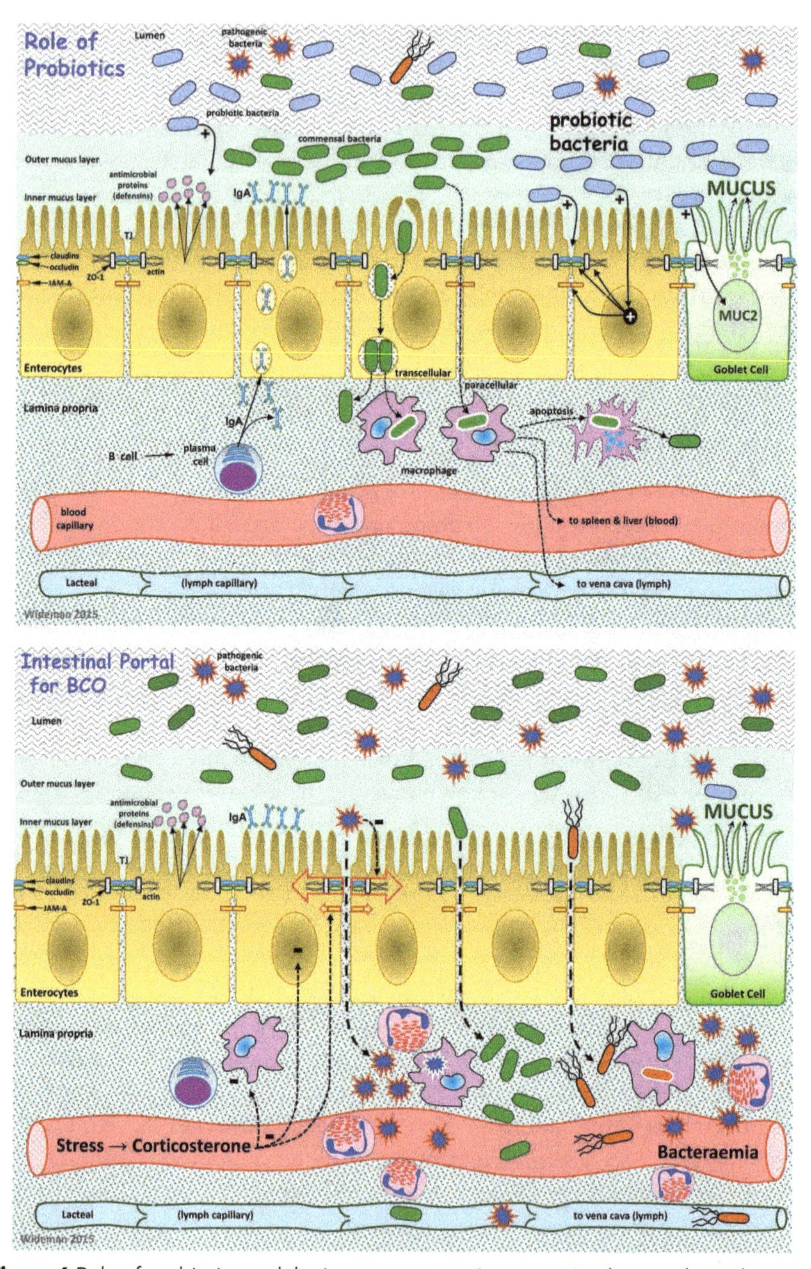

Figure 6 Role of probiotics and the immune system in preventing bacterial translocation across the healthy gastrointestinal tract (upper panel) and routes of bacterial translocation during dysbiosis (lower panel). Diagrams based on Steinwender at al. (2001), Ulluwishewa et al. (2011), Quinteiro-Filho et al. (2012a,b), Pastorelli et al. (2013) and Murugesan et al. (2014).

and consolidation of tight junctions into barriers capable of minimizing bacterial translocation via the paracellular route.

Factors known to influence the dynamic synthesis of new tight junction proteins and alter the integrity of existing tight junctions include stress and 'crosstalk' (direct cell to cell signalling) between gastrointestinal epithelial cells and commensal or pathogenic bacteria of the intestinal microbial community (Fig. 6, upper panel) (Steinwender et al., 2001; Ulluwishewa et al., 2011; Pastorelli et al., 2013). As shown in the lower panel of Fig. 6, stress hormones such as cortisol in mammals and corticosterone in birds suppress the immune system and can play a major role in creating 'leaky' tight junctions. Stress clearly facilitates the translocation of indigenous enteric bacteria, and chronic stress suppresses both the mucosal immune defences of the barrier epithelia and the innate immunity (Berg, 1992; Saunders et al., 1994; Ando et al., 2000).

Probiotics may reduce bacterial translocation by attenuating intestinal populations of pathogenic bacteria, by occupying binding sites on the brush border glycocalyx to prevent binding by pathogenic bacteria (competitive exclusion), by enhancing the integrity of existing tight junctions, by promoting gene expression and synthesis of tight junction proteins, by stimulating gene expression and secretion of protective mucus or by priming the immune system to produce a more robust immune exclusion barrier and to more effectively eliminate translocated bacteria (Fig. 6, lower panel) (Steinwender at al., 2001; Smirnov et al., 2005; Ulluwishewa et al., 2011; Pastorelli et al., 2013; Murugesan et al., 2014). Indeed, commensal and probiotic bacterial species that enhance intestinal barrier integrity by stimulating tight junction protein expression and the formation of occlusive tight junctional complexes are also effective in preventing bacterial translocation (Ulluwishewa et al., 2011; Pastorelli et al., 2013).

In broilers, heat stress and enhanced intestinal microbial challenges have been demonstrated to impair the integrity of tight junctions and facilitate bacterial translocation across the epithelium of the small intestine (Quinteiro-Filho et al., 2010, 2012a,b; Murugesan et al., 2014). It has also been demonstrated that probiotics alone or in combination with prebiotics can attenuate intestinal barrier dysfunction in broilers challenged by heat stress or pathogenic bacteria (Sohail et al., 2010, 2012; Murugesan et al., 2014; Song et al., 2014). For example, as shown in Fig. 7, the intestinal epithelium from broiler chicks reared on reused/dirty litter to simulate a typical commercial enteric challenge had a lower permeability to macromolecules (e.g. bacterial lipopolysaccharide), a higher transepithelial resistance and a higher mucin gene expression level when the feed contained a probiotic (Murugesan et al., 2014).

In view of concerns regarding the development of antibiotic resistance in bacteria commonly associated with osteomyelitis (McNamee and Smyth,

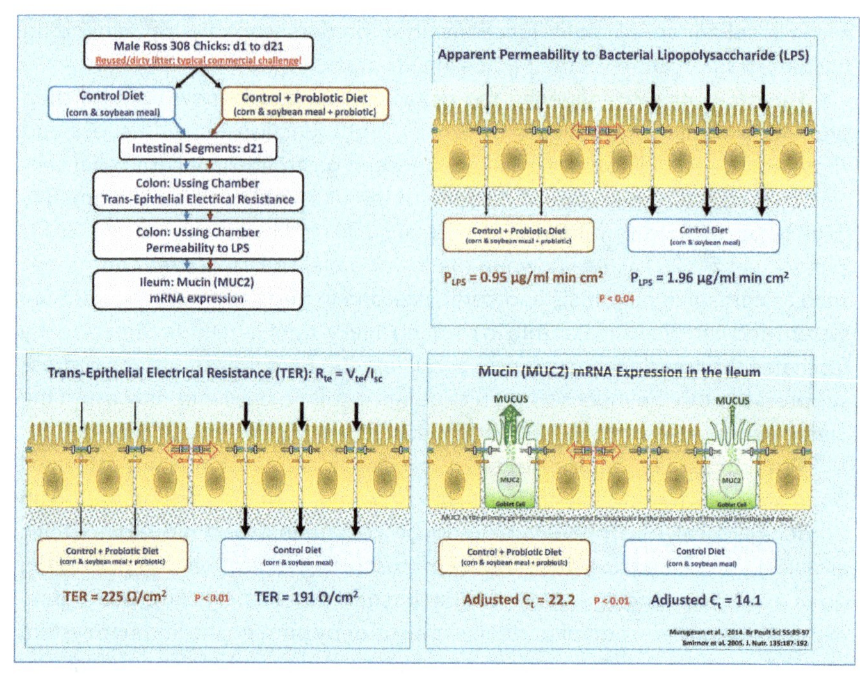

Figure 7 Broiler chicks reared on reused/dirty litter to simulate a typical commercial enteric challenge were fed a control diet or the same diet containing a probiotic. Intestinal segments were harvested on day 21, and Ussing chambers were used to assess intestinal permeability. Probiotic inclusion in the diet reduced the epithelial permeability to macromolecules (e.g. bacterial lipopolysaccharide) and increased the transepithelial resistance (reflecting improved tight junction barrier function). The probiotic also increased ileal mucin gene expression levels when compared with ileal samples from chicks fed the control diet (Murugesan et al., 2014).

2000; Waters et al., 2011), it was our hypothesis that probiotics potentially can provide a plausible alternative for reducing the incidence of BCO when the intestinal tract is the site of bacterial translocation. Five independent experiments conducted over the course of two years and using four different broiler lines demonstrated that prophylactically administering probiotics in the feed beginning at one day of age significantly reduced the incidence of lameness attributable to BCO (Wideman et al., 2012).

The first four of these experiments evaluated a proprietary probiotic containing *Enterococcus faecium*, *Bifidobacterium animalis*, *Pediococcus acidilactici* and *Lactobacillus reuteri*. The fifth experiment evaluated a proprietary single-microbe probiotic containing *E. faecium*. Prophylactically providing these probiotics in the feed consistently reduced the incidence of BCO lameness by at least 50% and without attenuating growth performance when compared with broilers that also were reared on wire flooring but were not provided probiotics in their feed (Wideman et al., 2012). In subsequent studies, a proprietary probiotic containing *Bacillus subtilis* significantly delayed the age of onset and reduced the cumulative incidence of BCO lameness in broilers reared on wire flooring, whereas experiments conducted with a different proprietary *B. subtilis* probiotic had no significant impact on the incidence of BCO lameness (Wideman et al., 2015a). Accordingly, although the specific biological mechanism remains to be determined, these experiments provide the first evidence that some, but not all, probiotics can significantly interrupt the pathogenesis of lameness attributable to BCO.

Trials conducted on accumulated litter in commercial broiler facilities have also demonstrated the practical efficacy of probiotics for reducing the incidence of BCO (Wideman, personal observations). The efficacy of feeding probiotics prophylactically to reduce the incidence of BCO implicates the gastrointestinal tract as a major avenue of bacterial translocation during the pathogenesis of this disease. It is our hypothesis that susceptibility to BCO lameness is minimized by probiotics that attenuate the translocation of pathogenic bacteria into the bloodstream. Based on recent evidence that the proximal femora and tibiae harbour complex microbial communities in their growth plates and metaphyses and that major differences exist in the microbial communities in different bones (femur versus tibia) and different lesion categories (macroscopically normal versus gross BCO lesions), it is intriguing to speculate that effective probiotics may improve bone health by modulating the composition and diversity of the microbial communities within the growth plates (Jiang et al., 2015). Probiotics potentially might influence growth plate microbial communities by reducing pathogen translocation, by enhancing the immune response to translocating microbial species or via translocation of the probiotic species to the growth plates followed by direct modulation of local microbial communities *in situ*.

6 Summary and conclusions

The current hypothesis for the pathogenesis of BCO proposes that exponential increases in early body weight impose excessive torque and shear stress on structurally immature epiphyseal and physeal growth plate cartilage, creating osteochondrotic wound sites that serve as ideal niches for colonization by haematogenously distributed opportunistic bacteria. We do not understand how pathogenic bacteria circulating in the blood avoid elimination by the immune system, or how complex microbial communities are allowed to persist in the growth plates of affected bones without being eliminated by an inflammatory response. Lameness is triggered when pathogenic bacteria create large abscesses primarily within the proximal ends of the femora and tibiae as well as in the flexible thoracic vertebrae.

Experimental models for triggering BCO are essential for confirming key triggering mechanisms, revealing innate susceptibilities and providing a reliable test bed for evaluating practical prophylactic and therapeutic strategies. Attempts to initiate BCO via a respiratory challenge have been only marginally successful, whereas osteomyelitis has repeatedly been reproduced by oral challenges and by injecting broilers and turkeys intravenously with appropriate strains of pathogenic bacteria. Intravenous injections of bacterial pathogens acutely overwhelm the body's defence mechanisms, initiate sustained septicaemia and tend to trigger widely dispersed osteomyelitis instead of the focal femoral, tibial and vertebral lesions that are pathognomonic for BCO. Intravenous injection models are unlikely to be useful for detecting differences in BCO susceptibility among broiler lines that may primarily depend on differences in innate susceptibility to osteochondrosis, nor are such models useful for demonstrating relative efficacies of practical prophylactic treatments that improve barrier integrity and thereby reduce bacterial translocation. Mechanical models based on wire flooring consistently trigger BCO by creating sustained footing instability that persistently exerts excessive mechanical stress on susceptible leg and vertebral joints. Mechanical models also act as chronic physiological stressors that promote indigenous bacterial translocation and proliferation attributed to stress-mediated immunosuppression. By consistently triggering significant incidences of BCO in research flocks, mechanical models have provided a reliable experimental protocol for comparing genetic lines and chick sources and for evaluating practical treatment strategies. Mechanical models avoid the need to purposefully inoculate birds with known pathogens based on the assumption that the offending pathogens already are present but quiescent within the bird or in the environment, waiting for an effective triggering scenario (e.g. stress, immunosuppression) and attainment of sufficient body mass to create a conducive wound site (e.g. osteochondrotic clefts, focal necrosis). Pathogenic bacteria may be transmitted from breeder parents, contaminated eggshells or hatchery sources or may enter the circulation via translocation through the integument, respiratory system or intestinal tract. Translocation of orally administered or indigenous enteric bacteria and haematogenous distribution by the circulatory system throughout the body has been demonstrated repeatedly. Most bacterial translocation across the intestinal barrier appears to utilize the paracellular route via 'leaky' tight junctions. Probiotics are known to improve tight junction integrity and reduce bacterial translocation. Furthermore, recent experiments provide the first evidence that some, but not all, probiotics can significantly interrupt the pathogenesis of lameness attributable to BCO. The efficacy of feeding probiotics prophylactically to reduce the incidence of BCO implicates the gastrointestinal tract as a major avenue of bacterial translocation during the pathogenesis of this disease. The roles of probiotics in improving the barrier function of the intestinal epithelium remain to be revealed. It is intriguing to speculate that probiotic species may translocate into the bloodstream to modulate internal microbial communities in various organs and tissues, including the growth plates of bones. Further research is necessary to study the role of probiotics in arresting the pathogenesis of BCO.

7 Where to look for further information

- Avian leg bone growth: Wideman, R. F., and R. D. Prisby. 2013. Bone circulatory disturbances in the development of spontaneous bacterial chondronecrosis with osteomyelitis: A translational model for the pathogenesis of femoral head necrosis. *Frontiers in Science (Front. Endocrin.)* 3: 183. doi:10.3389/fendo.2012.00183.
- Osteochondrosis review: Ytrehus, B., C. S. Carlson, and S. Ekman. 2007. Etiology and pathogenesis of osteochondrosis. *Vet. Pathol.* 44: 429-48.
- BCO review: McNamee, P. T., and J. A. Smyth. 2000. Bacterial chondronecrosis with osteomyelitis ('femoral head necrosis') of broiler chickens: A review. *Avian Pathol.* 29: 253-70.
- BCO review: Wideman, R. F. 2015. Bacterial chondronecrosis with osteomyelitis and lameness in broilers: A review. *Poult. Sci.* dx.doi.org/10.3382/ps/pev320.
- Intestinal tight junctions and probiotics: Ulluwishewa, D., R. C. Anderson, W. C. McNabb, P. J. Moughan, J. M. Wells, and N. C. Roy. 2011. Regulation of tight junction permeability by intestinal bacteria and dietary components. *J. Nutr.* 141: 769-76.
- Annual meetings: Symposium on Gut Health in Production of Food Animals (http://www.guthealthsymposium.com).

8 References

Alderson, M., and S. Nade. 1987. Natural history of acute septic arthritis in an avian model. *J. Orthopaed. Res.* 5: 261-74.

Alderson, M., K. Emslie, D. Speers, and S. Nade. 1986a. Transphyseal blood vessels exist in avian species. *J. Anat.* 146: 217-24.

Alderson, M., D. Speers, K. Emslie, and S. Nade. 1986b. Acute haematogenous osteomyelitis and septic arthritis - a single disease. *J. Bone Joint Surg.* 68B: 268-74.

Alfonso, M., and H. J. Barnes. 2006. Neonatal osteomyelitis associated with *Staphylococcus aureus* in turkey poults. *Avian Dis.* 50: 148-51.

Al-Rubaye, A. A., K. Estill, R. F. Wideman, and D. D. Rhoads. 2014. 16S rRNA-based diagnosis and whole-genome sequencing of bacteria cultured from lame broilers with osteomyelitis. *Poult. Sci.* 93 (E-Suppl. 1): 314P.

Ando, T., R. F. Brown, R. D. Berg, and A. J. Dunn. 2000. Bacterial translocation can increase plasma corticosterone and brain catecholamine and indoleamine metabolism. *Am. J. Physiol. Reg. Integr. Comp. Physiol.* 279: R2164–72.

Andreasen, J. R., C. B. Andreasen, M. Anwer, and A. E. Sonn. 1993. Heterophil chemotaxis in chickens with natural staphylococcal infections. *Avian Dis.* 37: 284-9.

Applegate, T. J., and M. S. Lilburn. 2002. Growth of the femur and tibia of a commercial broiler line. *Poult. Sci.* 81: 1289-94.

Bailey, J. S., N. A. Cox, D. E. Cosby, and L. J. Richardson. 2005. Movement and persistence of *Salmonella* in broiler chickens following oral or intracloacal inoculation. *J. Food Protect.* 68: 2698-2701.

Bar-Shira, E., I. Cohen, O. Elad, and A. Friedman. 2014. Role of goblet cells and mucin layer in protecting maternal IgA in precocious birds. *Develop. Comp. Immunol.* 44: 186-94.

Beaumont, G. D. 1967. The intraosseous vasculature of the ulna of *Gallus domesticus*. *J. Anat.* 101: 543-54.

Berg, R. D. 1992. Bacterial translocation from the gastrointestinal tract. *J. Med.* 23: 217-44.

Bohorquez, D. V., N. E. Bohorquez, and P. R. Ferket. 2011. Ultrastructural development of the small intestinal mucosa in the embryo and turkey poult: A light and electron microscopy study. *Poult. Sci.* 90: 842-55.

Bond, P. L., T. W. Sullivan, J. H. Douglas, and L. G. Robeson. 1991. Influence of age, sex, and method of rearing on tibia length and mineral deposition in broilers. *Poult. Sci.* 70: 1936-42.

Bradshaw, R. H., R. D. Kirkden, and D. M. Broom. 2002. A review of the aetiology and pathology of leg weakness in broilers in relation to welfare. *Avian Poult. Biol. Rev.* 13: 45-103.

Buhr, R. J., N. A. Cox, N. J. Stern, M. T. Musgrove, J. L. Wilson, and K. L. Hiett. 2002. Recovery of Campylobacter from segments of the reproductive tract of broiler breeder hens. *Avian Dis.* 46: 919-24.

Butterworth, A. 1999. Infectious components of broiler lameness: A review. *W. Poult. Sci. J.* 55: 327-52.

Byrne, C. M., M. Clyne, and B. Bourke. 2007. *Campylobacter jejuni* adhere to and invade chicken intestinal epithelial cells *in vitro*. *Microbiol.* 153: 561-9.

Carnaghan, R. B. A. 1966. Spinal cord compression due to spondylitis caused by *Staphylococcus pyogenes*. *J. Comp. Pathol.* 76: 9-14.

Carter, P. B., and F. M. Collins. 1974. The route of enteric infection in normal mice. *J. Exp. Med.* 139: 1189-1203.

Church, L. E., and L. C. Johnson. 1964. Growth of long bones in the chicken. Rates of growth in length and diameter of the humerus, tibia, and metatarsus. *Am. J. Anat.* 114: 521-38.

Clark, A. G., and D. H. Bueschkens. 1985. Laboratory infection of chicken eggs with *Campylobacter jejuni* by using temperature and pressure differentials. *Appl. Environ. Microbiol.* 49: 1467-71.

Clark, S., R. Porter, B. McComb, R. Lippert, S. Olsen, S. Nohner, and H. L. Shivaprasad. 2010. Clostridial dermatitis and cellulitis: An emerging disease in turkeys. *Avian Dis.* 54: 788-94.

Cossart, P., and P. J. Sansonetti. 2004. Bacterial invasion: The paradigms of enteroinvasive pathogens. *Science* 304: 242-8.

Cox, N. A., N. J. Stern, K. L. Hiett, and M. Berrang. 2002. Identification of a new source of Campylobacter contamination in poultry: Transmission from breeder hens to broiler chickens. *Avian Dis.* 46: 535–41.

Cox, N. A., C. L. Hofacre, J. S. Bailey, R. J. Buhr, J. L. Wilson, K. L. Hiett, L. J. Richardson, M. T. Musgrove, D. E. Cosby, J. D. Tankson, Y. L. Vizzier, P. F. Cray, L. E. Vaughn, P. S. Holt, and D. V. Bourassa. 2005a. Presence of *Campylobacter jejuni* in various organs one hour, one day, and one week following oral or intracloacal inoculations of broiler chicks. *Avian Dis.* 49: 155–8.

Cox, N. A., J. S. Bailey, L. J. Richardson, R. J. Buhr, D. E. Cosby, J. L. Wilson, K. L. Hiett, G. R. Siragusa, and D. V. Bourassa. 2005b. Presence of naturally occurring *Campylobacter* and *Salmonella* in the mature and immature ovarian follicles of late-life broiler breeder hens. *Avian Dis.* 49: 285–7.

Cox, N. A., L. J. Richardson, R. J. Buhr, P. J. Fedorka-Cray, J. S. Bailey, J. L. Wilson, and K. L. Hiett. 2006a. Natural presence of *Campylobacter* spp. in various internal organs of commercial broiler breeder hens. *Avian Dis.* 50: 450–3.

Cox, N. A., L. J. Richardson, R. J. Buhr, J. K. Northcutt, B. D. Fairchild, and J. M. Mauldin. 2006b. Presence of inoculated *Campylobacter* and *Salmonella* in unabsorbed yolks of male breeders raised as broilers. *Avian Dis.* 50: 430–3.

Cox, N. A., L. J. Richardson, R. J. Buhr, J. K. Northcutt, P. J. Fedorka-Cray, J. S. Bailey, B. D. Fairchild, and J. M. Mauldin. 2006c. Natural occurrence of *Campylobacter* species, *Salmonella* serovars, and other bacteria in unabsorbed yolks of market-age commercial broilers. *J. Appl. Poult. Res.* 15: 551–7.

Cox, N. A., L. J. Richardson, R. J. Buhr, J. K. Northcutt, J. S. Bailey, P. F. Cray, and K. L. Hiett. 2007. Recovery of *Campylobacter* and *Salmonella* serovars from the spleen, liver and gallbladder, and ceca of six- and eight-week-old commercial broilers. *J. Appl. Poult. Res.* 16: 477–80.

Daum, R. S., W. H. Davis, K. B. Farris, R. J. Campeau, D. M. Mulvihill, and S. M. Shane. 1990. A model of staphylococcus aureus bacteremia, septic arthritis, and osteomyelitis in chickens. *J. Orthopaed. Res.* 8: 804–13.

Desmidt, M., R. Ducatelle, and F. Haesebrouck. 1998. Serological and bacteriological observations on experimental infection with *Salmonella hadar* in chickens. *Vet. Microbiol.* 60: 259–69.

Devriese, L. A., A. H. Devos, and J. Beumer. 1972. *Staphylococcus aureus* colonization on poultry after experimental spray inoculations. *Avian Dis.* 16: 656–65.

Dinev, I. 2009. Clinical and morphological investigations on the prevalence of lameness associated with femoral head necrosis in broilers. *Br. Poult. Sci.* 50: 284–90.

Duff, S. R. I. 1984. Capital femoral epiphyseal infarction in skeletally immature broilers. *Res. Vet. Sci.* 37: 303–9.

Duff, S. R. I. 1990a. Do different forms of spondylolisthesis occur in broiler fowls? *Avian Pathol.* 19: 279–94.

Duff, S. R. I. 1990b. Diseases of the musculoskeletal system. In F. T. W. Jordan (ed.), *Poultry Diseases*, 3rd ed. (pp. 254–83). London, UK: Bailliere Tindall.

El-Lethey, H., B. Huber-Eicher, and T. W. Jungi. 2003. Exploration of stress-induced immunosuppression in chickens reveals both stress-resistant and stress-susceptible antigen responses. *Vet. Immunol. Immunopathol.* 95: 91–101.

Emslie, K. R., and S. Nade. 1983. Acute hematogenous staphylococcal osteomyelitis: A description of the natural history in an avian model. *Am. J. Pathol.* 110: 333–45.

Emslie, K. R., and S. Nade. 1985. Acute hematogenous staphylococcal osteomyelitis. *Comp. Pathol. Bulletin* 17: 2–3.

Emslie, K. R., N. R. Ozanne, and S. M. L. Nade. 1983. Acute hematogenous osteomyelitis: An experimental model. *J. Pathol.* 141: 157–67.

Emslie, K. R., L. M. Fenner, and S. M. L. Nade. 1984. Acute haematogenous osteomyelitis: II. The effect of a metaphyseal abscess on the surrounding blood supply. *J. Pathol.* 142: 129–34.

Funkhouser, L. J., and S. R. Bordenstein. 2013. Mom knows best: The universality of maternal microbial transmission. *PLoS Biol.* 11(8): e1001631. doi::10.1371/journal.pbio.1001631.

Gilley, A. D., H. Lester, I. Y. Pevzner, N. B. Anthony, and R. F. Wideman, Jr. 2014. Evaluating portable wire flooring models for inducing bacterial chondronecrosis with osteomyelitis (BCO) in broilers. *Poult. Sci.* 93: 1354–67.

Griffiths, G. L., W. L. Hopkinson, and J. Lloyd. 1984. Staphylococcal necrosis in the head of the femur in broiler chickens. *Austral. Vet. J.* 61: 293.

Hales, S. 1727. *Static Essays, Vol I: Vegetable Staticks*. London: W. Innys & Woodward.

Hand, W. L., N. L. King-Thompson, and J. D. Johnson. 1984. Influence of bacterial-antibiotic interactions on subsequent antimicrobial activity of alveolar macrophages. *J. Infect. Dis.* 149: 271–6.

Havenstein, G. B., P. R. Ferket, S. E. Scheidler, and B. T. Larson. 1994. Growth, livability and feed conversion of 1957 vs. 1991 broilers when fed 'typical' 1957 and 1991 broiler diets. *Poult. Sci.* 73: 1785–94.

Haye, U., and P. C. M. Simons. 1978. Twisted legs in broilers. *Br. Poult. Sci.* 19: 549–57.

Howlett, C. R. 1980. The fine structure of the proximal growth plate and metaphysis of the avian tibia: Endochondral osteogenesis. *J. Anat.* 130: 745–68.

Howlett, C. R., M. Dickson, and A. K. Sheridan. 1984. The fine structure of the proximal growth plate of the avian tibia: Vascular supply. *J. Anat.* 139: 115–32.

Huff, G. R., W. E. Huff, J. M. Balog, and N. C. Rath. 1998. Effects of dexamethasone immunosuppression on turkey osteomyelitis complex in an experimental *Escherichia coli* respiratory infection. *Poult. Sci.* 77: 654–61.

Huff, G. R., W. E. Huff, J. M. Balog, and N. C. Rath. 1999. Sex differences in the resistance of turkeys to *Escherichia coli* challenge after immunosuppression with dexamethasone. *Poult. Sci.* 78: 38–44.

Huff, G. R., W. E. Huff, N. C. Rath, and J. M. Balog. 2000. Turkey osteomyelitis complex. *Poult. Sci.* 79: 1050–6.

Huff, G. R., W. E. Huff, J. M. Balog, N. C. Rath, N. B. Anthony, and K. E. Nestor. 2005. Stress response differences and disease susceptibility reflected by heterophil to lymphocyte ratio in turkeys selected for increased body weight. *Poult. Sci.* 84: 709–17.

Huff, G. R., W. E. Huff, N. C. Rath, and J. M. Balog, N. B. Anthony, and K. E. Nestor. 2006. Stress-induced colibacillosis and turkey osteomyelitis complex in turkeys selected for increased body weight. *Poult. Sci.* 85: 266–72.

Humphrey, T. J., and D. G. Lanning. 1988. The vertical transmission of salmonellas and formic acid treatment of chicken feed. *Epidem. Inf.* 100: 43–9.

Hunt, C. D., D. A. Ollerich, and F. H. Nielsen. 1979. Morphology of the perforating cartilage canals in the proximal tibial growth plate of the chick. *Anat. Rec.* 194: 143–57.

Jensen, M. M., W. C. Downs, J. D. Morrey, T. R. Nicoll, S. D. Lefevre, and C. M. Meyers. 1987. Staphylococcosis of turkeys. 1. Portal of entry and tissue colonisation. *Avian Dis.* 31: 64–9.

Jiang, T., R. K. Mandal, R. F. Wideman Jr., A. Khatiwara, I. Pevzner, and Y. M. Kwon. 2015. Molecular survey of bacterial communities associated with bacterial chondronecrosis with osteomyelitis in broilers. *PLoS ONE*, doi:10.1371/journal.pone.0124403, 16 April 2015. http://www.ncbi.nlm.nih.gov/pubmed/25881241.

Joiner, K. S., F. J. Hoerr, E. van Santen, and S. J. Ewald. 2005. The avian major histocompatibility complex influences bacterial skeletal disease in broiler breeder chickens. *Vet. Pathol.* 42: 275–81.

Julian, R. J. 1985. Osteochondrosis, dyschondroplasia, and osteomyelitis causing femoral head necrosis in turkeys. *Avian Dis.* 29: 854–66.

Julian, R. J. 2005. Production and growth related disorders and other metabolic diseases of poultry – A review. *Vet. J.* 169: 350–69.

Khan, M. A., N. O. Olson, and D. O. Overman. 1977. Spontaneous spondylolisthesis in embryonic and adult chick. *Poult. Sci.* 56: 689–97.

Kember, N. F., and J. K. Kirkwood. 1987. Cell kinetics and longitudinal bone growth in birds. *Cell Tissue Kinet.* 20: 625–9.

Kember, N. F., J. K. Kirkwood, P. J. Duignan, D. Godfrey, and D. J. Spratt. 1990. Comparative cell kinetics of avian growth plates. *Res. Vet. Sci.* 49: 283–8.

Kense, M. J., and W. J. M. Landman. 2011. *Enterococcus cecorum* infections in broiler breeders and their offspring: Molecular epidemiology. *Avian Pathol.* 40: 603–12.

Kestin, S. C., G. Su, and P. Sorensen. 1999. Different commercial broiler crosses have different susceptibilities to leg weakness. *Poult. Sci.* 78: 1085–90.

Kestin, S. C., S. Gordon, G. Su, and P. Sorensen. 2001. Relationship in broiler chickens between lameness, live weight, growth rate and age. *Vet. Rec.* 148: 195–7.

Kibenge, F. S. B., G. E. Wilcox, and D. A. Pass. 1983. Pathogenicity of four strains of *Staphylococcus aureus* isolated from chickens with clinical tenosynovitis. *Avian Pathol.* 12: 213–20.

Kirkwood, J. K., P. J. Duignan, N. F. Kember, P. M. Bennett, and D. J. Price. 1989a. The growth of the tarsometatarsus bone in birds. *J. Zool., Lond.* 217: 403–16.

Kirkwood, J. K., D. M. J. Spratt, and P. J. Duignan. 1989b. Patterns of cell proliferation and growth rate in limb bones of the domestic fowl (*Gallus domesticus*). *Res. Vet. Sci.* 47: 139–47.

Kremer, C. J., K. M. O'Mera, S. L. Layton, B. M. Hargis, and K. Cole. 2011. Evaluation of recombinant *Salmonella* expressing the flagellar protein flic for persistence and enhanced antibody response in commercial turkeys. *Poult. Sci.* 90: 752–8.

Kuhlers, D. L., and G. R. McDaniel. 1996. Estimates of heritabilities and genetic correlations between tibial dyschondroplasia expression and body weight at two ages in broilers. *Poult. Sci.* 75: 959–61.

Leach, R. M., and M. C. Nesheim. 1965. Nutritional, genetic and morphological studies of an abnormal cartilage formation in young chicks. *J. Nutr.* 86: 236–44.

Leach, R. M. Jr., and C. V. Gay. 1987. Role of epiphyseal cartilage in endochondral bone formation. *J. Nutr.* 117: 784–90.

Liljebjelke, K. A., C. L. Hofacre, T. Liu, D. G. White, S. Ayers, S. Young, and J. J. Maurer. 2005. Vertical and horizontal transmission of Salmonella within integrated broiler production system. *Foodborne Pathog. Dis.* 2: 90–102.

Lissner, C. R., R. N. Swanson, and A. D. O'Brien. 1983. Genetic control of the innate resistance of mice to *Salmonella typhimurium*: Expression of the Ity gene in peritoneal and splenic macrophages isolated *in vitro*. *J. Immunol.* 131: 3003–13.

Lutfi, A. M. 1970a. Study of cell multiplication in the cartilagenous upper end of the tibia of the domestic fowl by tritiated thymidine autoradiography. *Acta Anat.* 76: 454–63.

Lutfi, A. M. 1970b. The mode of growth, fate and function of cartilage canals. *J. Anat.* 106: 135–45.

Martin, L. T., M. P. Martin, and H. J. Barnes. 2011. Experimental reproduction of enterococcal spondylitis in male broiler breeder chickens. *Avian Dis.* 55: 273–8.

McCaskey, P. C., G. N. Rowland, R. K. Page, and L. R. Minear. 1982. Focal failures of endochondral ossification in the broiler. *Avian Dis.* 26: 701–17.

McCullagh, J. J., P. T. McNamee, J. A. Smyth, and H. J. Ball. 1998. The use of pulsed-field gel electrophoresis to investigate the epidemiology of *Staphylococcus aureus* infection in commercial broiler flocks. *Vet. Microbiol.* 63: 275–81.

McNamee, P. T., and J. A. Smyth. 2000. Bacterial chondronecrosis with osteomyelitis ('femoral head necrosis') of broiler chickens: A review. *Avian Pathol.* 29: 253–70.

McNamee, P. T., J. J. McCullagh, B. H. Thorp, H. J. Ball, D. Graham, S. J. McCullough, D. McConaghy, and J. A. Smyth. 1998. Study of leg weakness in two commercial broiler flocks. *Vet. Rec.* 143: 131–5.

McNamee, P. T., J. J. McCullagh, J. D. Rodgers, B. H. Thorp, H. J. Ball, T. J. Connor, D. McConaghy, and J. A. Smyth. 1999. Development of an experimental model of bacterial chondronecrosis with osteomyelitis in broilers following exposure to Staphlococcus aureus by aerosol, and inoculation with chicken anemia and infectious bursal disease viruses. *Avian Pathol.* 28: 26–35.

Mercer, J. T., and W. G. Hill. 1984. Estimation of genetic parameters for skeletal defects in broiler chickens. *Heredity* 53: 193–203.

Murugesan, G. R., N. K. Gabler, and M. E. Persia. 2014. Effects of direct-fed microbial supplementation on broiler performance, intestinal nutrient transport and integrity under experimental conditions with increased microbial challenge. *Br. Poult. Sci.* 55: 89–97.

Mutalib, A., C. Riddell, and A. D. Osborne. 1983a. Studies on the pathogenesis of staphylococcal osteomyelitis in chickens. I. Effect of stress on experimentally induced osteomyelitis. *Avian Dis.* 27: 141–56.

Mutalib, A., C. Riddell, and A. D. Osborne. 1983b. Studies on the pathogenesis of staphylococcal osteomyelitis in chickens. II. Role of the respiratory tract as a route of infection. *Avian Dis.* 27: 157–60.

Naas, I. A., I. C. L. A. Paz, M. S. Baracho, A. G. Menezes, L. G. F. Bueno, I. C. L. Almeida, and D. J. Moura. 2009. Impact of lameness on broiler well-being. *J. Appl. Poult. Res.* 18: 432–9.

Nairn, M. E. (1973). Bacterial osteomyelitis and synovitis in the turkey. *Avian Dis.* 17: 504–17.

Nairn, M. E., and A. R. A. Watson. 1972. Leg weakness of poultry – a clinical and pathological characterisation. Aust. Vet. J. 48: 645–56.

Nehme, P. A., E. K. Barbour, V. K. Sagherian, S. K. Hamadeh, and R. K. Zurayk. 2005. Baseline data on enumerated tracheal bacterial flora and drug susceptibility in chickens reared under different systems. *Inter. J. Appl. Res. Vet. Med.* 3: 372–7.

Nestor, K. E. 1984. Genetics of growth and reproduction in the turkey. 9. Long-term selection for increased 16-week body weight. *Poult. Sci.* 63: 2114–22.

Nestor, K. E., and J. W. Anderson. 1998. Effect of crossing a line selected for increased shank width with two commercial sire lines on performance and walking ability of turkeys. *Poult. Sci.* 77: 1601–7.

Nicoll, T. R., and M. M. Jensen. 1986. Preliminary studies on bacterial interference of staphylococcosis of chickens. *Avian Dis.* 31: 140–4.

Pastorelli, L., C. De Salvo, J. R. Mercado, M. Vecchi, and T. T. Pizarro. 2013. Central role of the gut epithelial barrier in the pathogenesis of chronic intestinal inflammation: Lessons learned from animal models and human genetics. *Frontiers in Science (Front. Immunol.)* 4: 280. doi:10.3389/fimmu.2013.002080.

Pearson, A. D., M. H. Greenwood, R. K. A. Feltham, T. D. Healing, J. Donaldson, D. M. Jones, and R. R. Colwell. 1996. Microbial ecology of *Campylobacter jejuni* in a United Kingdom supply chain: Intermittent common source, vertical transmission, and amplification by flock propagation. *Appl. Environ. Microbiol.* 62: 4614–20.

Prisby, R., T. Menezes, J. Campbell, T. Benson, E. Samraj, I. Pevzner, and R. F. Wideman, Jr. 2014. Kinetic examination of femoral bone modeling in broilers. *Poult. Sci.* 93: 1122–9.

Quinteiro-Filho, W. M., A. Ribeiro, V. Ferraz-de-Paula, M. L. Pinheiro, M. Sakai, L. R. M. Sa, A. J. P. Ferreira, and J. Palermo-Neto. 2010. Heat stress impairs performance parameters, induces intestinal injury, and decreases macrophage activity in broiler chickens. *Poult. Sci.* 89: 1905–14.

Quinteiro-Filho, W. M., M. V. Rodrigues, A. Ribeiro, V. Ferraz-de-Paula, M. L. Pinheiro, L. R. M. Sa, A. J. P. Ferreira, and J. Palermo-Neto. 2012a. Acute heat stress impairs performance parameters and induces mild intestinal enteritis in broiler chickens: Role of acute hypothalamic-pituitary-adrenal axis of activation. *J. Anim. Sci.* 90: 1986–94.

Quinteiro-Filho, W. M., A. V. S. Gomes, M. L. Pinheiro, A. Ribeiro, V. Ferraz-de-Paula, C. S. Astolfi-Ferreira, A. J. P. Ferreira, and J. Palermo-Neto. 2012b. Heat stress impairs performance and induces intestinal inflammation in broiler chickens infected with *Salmonella* Enteritidis. *Avian Pathol.* 41: 421–7.

Reiber, M. A., J. A. McInroy, and D. E. Conner. 1995. Enumeration and identification of bacteria in chicken semen. *Poult. Sci.* 74: 795–9.

Richardson, L. J., N. A. Cox, R. J. Buhr, and M. A. Harrison. 2011. Isolation of *Campylobacter* from circulating blood of commercial broilers. *Avian Dis.* 55: 375–8.

Riddell, C. 1973. Studies on spondylolisthesis (kinky-back) in broiler chickens. *Avian Pathol.* 2: 295–304.

Riddell, C. 1975. Studies on the pathogenesis of tibial dyschondroplasia in chickens. II. Growth rate of long bones. *Avian Dis.* 19: 490–6.

Riddell, C. 1976. Selection of broiler chickens for high and low incidence of tibial dyschondroplasia with observations on spondylolisthesis and twisted legs (perosis). *Poult. Sci.* 55: 145–51.

Riddell, C. 1983. Pathology of the skeleton and tendons of broiler chickens reared to roaster weights. I. crippled chickens. *Avian Dis.* 27: 950–62.

Riddell, C., M. W. King, and K. R. Gunasekera. 1983. Pathology of the skeleton and tendons of broiler chickens reared to roaster weights. II. Normal chickens. *Avian Dis.* 27: 980–91.

Rodgers, J. D., J. J. McCullagh, P. T. McNamee, J. A. Smyth, and H. J. Ball. 1999. Comparison of *Staphylococcus aureus* recovered from personnel in a poultry hatchery and in broiler parent farms with those isolated from skeletal disease in broilers. *Vet. Microbiol.* 69: 189–98.

Sandilands, V., S. Brocklehurst, N. Sparks, L. Baker, R. McGovern, B. Thorp, and D. Pearson. 2011. Assessing leg health in chickens using a force plate and gait scoring: How many birds is enough? *Vet. Rec.* 168: 77–83.

Saunders, P. R., U. Kosecka, D. M. McKay, and M. H. Perdue. 1994. Acute stressors stimulate ion secretion and increase epithelial permeability in rat intestine. *Am. J. Physiol.* 267: G794–9.

Shabbir, M. Z., T. Malys, Y. V. Ivanov, J. Park, M. A. B. Shabbir, M. Rabbani, T. Yaqub, and E. T. Harvill. 2015. Microbial communities present in the lower respiratory tract of clinically healthy birds in Pakistan. *Poult. Sci.* 94: 612–20.

Sheridan, A. K., C. R. Howlett, and R. W. Burton. 1978. The inheritance of tibial dyschondroplasia in broilers. *Br. Poult. Sci.* 19: 491–9.

Skeeles, K. J. 1997. Staphylococcosis. In B. W. Calnek, H. J. Barnes, C. W. Beard, L. R. McDougald, and Y. M. Saif (ed.), *Diseases of Poultry 10th Edition*, ch. 11 (pp. 247–53). Ames, IA, USA: Iowa State University Press.

Smeltzer, M. S., and A. F. Gillaspy. 2000. Molecular pathogenesis of staphylococcal osteomyelitis. *Poult. Sci.* 79: 1042–9.

Smirnov, A., R. Perez, E. Amit-Romach, D. Sklan, and Z. Uni. 2005. Mucin dynamics and microbial populations in chicken small intestine are changed by dietary probiotic and antibiotic growth promoter supplementation. *J. Nutr.* 135: 187–92.

Smith, H. W. 1954. Experimental staphylococcal infection in chickens. *J. Pathol. Bacteriol.* 67: 81–7.

Sohail, M. U., A. Ijaz, M. S. Yousaf, K. Ashraf, H. Zaneb, M. Aleem, and H. Rehman. 2010. Alleviation of cyclic heat stress in broilers by dietary supplementation of mannan-oligosaccharide and *Lactobacillus*-based probiotic: Dynamics of cortisol, thyroid hormones, cholesterol, C-reactive protein, and humoral immunity. *Poult. Sci.* 89: 1934–8.

Sohail, M. U., M. E. Hume, J. A. Byrd, D. J. Nisbet, A. Ijaz, A. Sohail, M. Z. Shabbir, and H. Rehman. 2012. Effect of supplementation of prebiotic mannan-oligosaccharides and probiotic mixture on growth performance of broilers subjected to chronic heat stress. *Poult. Sci.* 91: 2235–40.

Sohail, M. U., M. E. Hume, J. A. Byrd, D. J. Nisbet, M. Z. Shabbir, A. Ijaz, and H. Rehman. 2015. Molecular analysis of the caecal and tracheal microbiome of heat stressed broilers supplemented with prebiotic and probiotic. *Avian Pathol.* 44: 67–74.

Somes, R. G., Jr. 1969. Genetic perosis in the domestic fowl. *J. Hered.* 60: 163–6.

Song, J., K. Xiao, Y. L. Ke, L. F. Jiao, C. H. Hu, Q. Y. Diao, B. Shi, and X. T. Zou. 2014. Effect of a probiotic mixture on intestinal microflora, morphology, and barrier integrity of broilers subjected to heat stress. *Poult. Sci.* 93: 581–8.

Sorensen, P. 1992. The genetics of leg disorders. In C. C. Whitehead (ed.), *Bone Biology and Skeletal Disorders in Poultry* (pp. 213–29). Abingdon, United Kingdom: Carfax Publishing Company.

Soriani, M., I. Santi, A. Taddei, R. Rappuoli, G. Grandi, and J. L. Tedford. 2006. Group B *Streptococcus* crosses human epithelial cells by a paracellular route. *J. Infect. Dis.* 193: 241–50.

Speers, D. J., and S. M. L. Nade. 1985. Ultrastructural studies of adherence of Staphylococcus aureus in experimental acute hematogenous osteomyelitis. *Infection and Immunity* 49: 443-6.

Stalker, M. J., M. L. Brash, A. Weisz, R. M. Ouckama, and D. Slavic. 2010. Arthritis and osteomyelitis associated with *Enterococcus cecorum* infection in broiler and broiler breeder chickens in Ontario, Canada. *J. Vet. Diagn. Invest.* 22: 643-5.

Steinwender, G., G. Schimpl, B. Sixl, and H. H. Wenzl. 2001. Gut-derived bone infection in neonatal rat. *Pediatr. Res.* 50: 767-71.

Tate, C. R., W. C. Mitchell, and R. G. Miller. 1993. *Staphylococcus hyicus* associated with turkey stifle joint osteomyelitis. *Avian Dis.* 37: 905-7.

Thorp, B. H. 1988a. Vascular pattern of the developing knee joint in the domestic fowl. *Res. Vet. Sci.* 44: 89-99.

Thorp, B. H. 1988b. Vascular pattern of the developing intertarsal joint in the domestic fowl. *Res. Vet. Sci.* 44: 100-11.

Thorp, B. H. 1988c. Relationship between the rate of longitudinal bone growth and physeal thickness in the growing fowl. *Res. Vet. Sci.* 45: 83-5.

Thorp, B. H. 1994. Skeletal disorders in the fowl: A review. *Avian Pathol.* 23: 203-36.

Thorp, B. H., and D. Waddington. 1997. Relationships between the bone pathologies, ash and mineral content of long bones in 35-day-old broiler chickens. *Res. Vet. Sci.* 62: 67-73.

Thorp, B. H., C. C. Whitehead, L. Dick, J. M. Bradbury, R. C. Jones, and A. Wood. 1993. Proximal femoral degeneration in growing broiler fowl. *Avian Pathol.* 22: 325-42.

Ulluwishewa, D., R. C. Anderson, W. C. McNabb, P. J. Moughan, J. M. Wells, and N. C. Roy. 2011. Regulation of tight junction permeability by intestinal bacteria and dietary components. *J. Nutr.* 141: 769-76.

Uni, Z., Y. Noy, and D. Sklan. 1995. Post hatch changes in morphology and function of the small intestines in heavy- and light-strain chicks. *Poult. Sci.* 74: 1622-9.

Walser, M. M., F. L. Cherms, and H. E. Dziuk. 1982. Osseous development and tibial dyschondroplasia in five lines of turkeys. *Avian Dis.* 26: 265-70.

Waters, A. E., T. Contente-Cuomo, J. Buchhagen, C. M. Liu, L. Watson, K. Pierce, J. T. Foster, J. Bowers, E. M. Driebe, D. M. Engelthaler, P. S. Keim, and L. B. Price. 2011. Multi-drug resistant *Staphylococcus aureus* in US meat and poultry. *Clin. Infect. Dis.* 52: 1227-30.

Wideman, R. F. Jr. 2014. Bacterial chondronecrosis with osteomyelitis and lameness in broilers: Pathogenesis, experimental models, and preventative treatments. In W. H. A. Abdelrahman (ed.), *Probiotics in Poultry Production, Concept and Applications* (pp. 91-128). Sheffield, UK: 5m Publishing Ltd.

Wideman, R. F., and I. Pevzner. 2012. Dexamethasone triggers lameness associated with necrosis of the proximal tibial head and proximal femoral head in broilers. *Poult. Sci.* 91: 2464-74.

Wideman, R. F., and R. D. Prisby. 2013. Bone circulatory disturbances in the development of spontaneous bacterial chondronecrosis with osteomyelitis: a translational model for the pathogenesis of femoral head necrosis. *Frontiers in Science (Front. Endocrin.)* 3: 183. doi:10.3389/fendo.2012.00183.

Wideman, R. F., K. R. Hamal, J. M. Stark, J. Blankenship, H. Lester, K. N. Mitchell, G. Lorenzoni, and I. Pevzner. 2012. A wire flooring model for inducing lameness in broilers: Evaluation of probiotics as a prophylactic treatment. *Poult Sci.* 91: 870-83.

Wideman, R. F. Jr., A. Al-Rubaye, A. Gilley, D. Reynolds, H. Lester, D. Yoho, J. D. Hughes Jr., and I. Y. Pevzner. 2013. Susceptibility of four commercial broiler crosses to lameness attributable to bacterial chondronecrosis with osteomyelitis. *Poult. Sci.* 92: 2311-25.

Wideman, R. F. Jr., A. Al-Rubaye, D. Reynolds, D. Yoho, H. Lester, C. Spencer, J. M. Hughes, and I. Y. Pevzner. 2014. Bacterial chondronecrosis with osteomyelitis (BCO) in broilers: Influence of sires and straight-run vs. sex-separate rearing. *Poult. Sci.* 93: 1675-87.

Wideman, R. F. Jr., A. Al-Rubaye, Y. M. Kwon, J. Blankenship, H. Lester, K. N. Mitchell, I. Y. Pevzner, T. Lohrmann, and J. Schleifer. 2015a. Prophylactic administration of a combined prebiotic and probiotic, or therapeutic administration of enrofloxacin, to reduce the incidence of bacterial chondronecrosis with osteomyelitis in broilers. *Poult. Sci.* 94: 25-36.

Wideman, R. F. Jr., J. Blankenship, I. Y. Pevzner, and B. J. Turner. 2015b. Efficacy of 25-OH vitamin D_3 prophylactic administration for reducing lameness in broilers grown on wire flooring. *Poult. Sci.* 94: 1821-7.

Wijetunge, D. S., P. Dunn, E. Waller-Pendleton, V. Lintner, H. Lu, and S. Kariyawasam. 2012. Fingerprinting of poultry isolates of *Enterococcus cecorum* using three molecular typing methods. *J. Vet. Diag. Invest.* 24: 1166-71.

Williams, B., S. Solomon, D. Waddington, B. Thorp, and C. Farquharson. 2000. Skeletal development in meat-type chickens. *Br. Poult. Sci.* 41: 141-9.

Wise, D. R. 1970a. Carcass conformation comparisons of growing broilers and laying strain chickens. *Br. Poult. Sci.* 11: 325-32.

Wise, D. R. 1970b. Spondylolisthesis ('kinky back') in broiler chickens. *Res. Vet. Sci.* 11: 447-51.

Wise, D. R. 1971. Staphylococcal osteomyelitis of the avian vertebral column. *Res. Vet. Sci.* 12: 169-71.

Wise, D. R. 1973. The incidence and aetiology of avian spondylolisthesis (kinky-back). *Res. Vet. Sci.* 14: 1-10.

Wyers, M., Y. Cherel, and G. Plassiart. 1991. Late clinical expression of lameness related to associated osteomyelitis and tibial dyschondroplasia in male breeding turkeys. *Avian Dis.* 35: 408-14.

Yair, R., Z. Uni, and R. Shahar. 2012. Bone characteristics of late-term embryonic and hatchling broilers: Bone development under extreme growth rate. *Poult. Sci.* 91: 2614-20.

Yalcin, S., S. Ozkan, E. Coskuner, G. Bilgen, Y. Delen, Y. Kurtulmus, and T. Tanyalcin. 2001. Effects of strain, maternal age and sex on morphological characteristics and composition of tibial bone in broilers. *Br. Poult. Sci.* 42: 184-90.

Zhang, X., G. R. McDaniel, Z. S. Yalcin, and D. L. Kuhlers. 1995. Genetic correlations of tibial dyschondroplasia incidence with carcass traits in broilers. *Poult. Sci.* 74: 910-15.

Chapter 3

Bone health and associated problems in layer hens

Christina Rufener, University of California-Davis, USA; and Michael J. Toscano, University of Bern, Switzerland

1 Introduction

Housing and management systems for laying hens are variable but must allow for appropriate skeletal and cognitive development during rearing to ensure adult hens can support normal biological functioning while producing a large number of eggs. High egg production requires mobilization of resources including minerals and energy. Likely as a consequence of this mobilization and related factors, laying hens are believed to have weakened skeletal systems leading to specific problems including fractured keels during the laying period and other bone injuries during removal from the barn at the end of lay. To combat these problems, multiple strategies have been undertaken including effective breeding and nutritional strategies as well as adaptations of rearing practices and layer housing. The following chapter provides a summary of the basic skeletal system and its development, specific problems of bone health and efforts to reduce the problem. References to specific types of housing systems will be based on the three broad categories described elsewhere in this book. Within cage-free systems, housing type is further distinguished as single-tier or multi-tier (e.g. aviary) systems.

http://dx.doi.org/10.19103/AS.2020.0078.16

2 Bone development, growth and remodelling

Bone development, growth and remodelling of the avian skeleton are highly complex processes - especially as birds have special features including pneumatic (hollow) bones enabling flight and structures such as the keel bone serving as attachment point for the wing muscles (Sisson et al., 1975). To discuss influencing factors and potential solutions for bone health problems in adult laying hens, it is crucial to understand the mechanisms involved in the development and growth of the skeleton in pullets as well as ongoing bone remodelling processes after the onset of lay.

2.1 Bone development during rearing

The genesis of structures for cartilage formation is considered to be the earliest stage of skeletal development (Pines and Reshef, 2014) and starts as early as the first day of embryo development (Pourquié, 2004). At hatch and then continuing for the first 17 weeks of life, the structural skeleton becomes fully developed in terms of shape and dimensions, though it is likely to continue undergoing maturation (e.g. strength of internal linkages) until 35 weeks of age (Rath et al., 2000). Depending on the bone type, bone development and ossification are driven by either endochondral or intramembranous processes (Gilbert, 2000). Long bones of the appendicular skeleton such as the tibia or humerus develop by endochondral ossification, where cartilage is replaced by bone tissue. Flat bones such as the skull, on the other hand, are formed directly from connective tissue (mesenchyme) without a cartilage template. The keel bone is a special hybrid of these conditions as it is a rather flat bone (and therefore subject to intramembranous ossification), but has a cartilage template which is progressively ossifying with increasing age. Although the keel has previously been shown to be fully ossified at 40 weeks of age (Buckner et al., 1949), anecdotal reports have found that the keel tip (i.e. the most caudal part) remains mostly cartilaginous until 70 weeks of age.

Skeletal growth of long bones involves two main processes: elongation (i.e. longitudinal growth) and widening of the bones (Whitehead, 2004). Bones grow in length by proliferation (i.e. cellular division), maturation and calcification of cartilage cells at the growth plate located at the ends of the bone. Osteoblasts, the cells responsible for bone formation, then secrete bone tissue on the calcified cartilage to ossify the newly formed bone matrix. The second process of bone growth, widening of bones, involves the resorption of old bone in the medullary cavity by osteoclasts. Osteoblasts deposit new bone tissue on the outer surface of the bone, increasing the diameter of the bone and the medullary cavity. Once the pullet skeleton is fully developed, avian long bones are similar in structure to mammalian bones with an outer, compact cortical and an inner trabecular (or cancellous) bone (Fig. 1). Although both cortical and trabecular bones consist of the lamellar bone, the organization of lamellae differs and affects the function and stability of the bone. Cortical bone with a compact organization of lamellae serves as the principal structure of the skeleton, whereas trabecular bone arranged in rods and plates reduces bone density and allows long bones to compress if stress is applied.

2.2 Medullary bone and bone remodelling during lay

At the age of sexual maturity (17–19 weeks of age) and shortly before the onset of lay, the bone biology of laying hens changes dramatically. To meet the calcium demand for egg shell formation, avian females produce a special bone type referred to as medullary bone (Dacke et al., 1993). In contrast to the outer, compact cortical bone and the inner trabecular bone commonly found in all mammals, medullary bone is a woven, unorganized and highly calcified bone type that serves as a labile calcium source to cover high calcium requirements emerging during egg shell formation (Fig. 1; Etches, 1987). Once hens reach sexual maturity, the formation of structural bone, that is, the cortical and trabecular bones, ceases (Hudson et al., 1993) and the medullary bone begins to be deposited in the medullary cavities of long bones (Whitehead, 2004). The shift from the formation of structural bone towards the deposition of the medullary bone is driven by oestrogen, which increases osteoblastic activity and stimulates the formation of the woven bone instead of the lamellar bone (Whitehead and Fleming, 2000).

The medullary bone undergoes constant remodelling within the daily routine of bone depletion during egg shell calcification alternating with bone formation during periods when no egg shell is being produced (Kerschnitzki

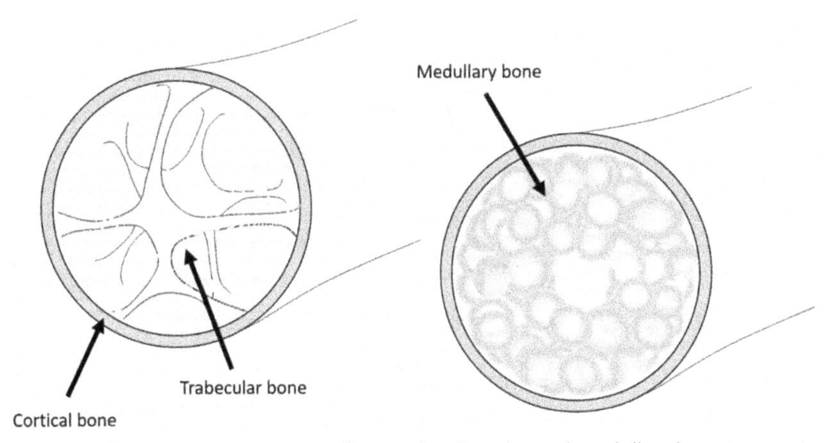

Figure 1 Schematic representation of cortical, trabecular and medullary bone in an avian long bone. Source: C. Rufener.

et al., 2014; Johnson, 2015). No new structural bone is formed while the hen is in lay, and bone resorption by osteoclastic is not specific for one bone type (Fleming et al., 2006). As a result, calcium is mobilized during egg laying from both medullary and structural bone, but with different rates (Simkiss, 1967) as the medullary bone can be metabolized 10–15 times faster than the cortical bone (Hurwitz, 1965). Junglefowl - the ancestor of all commercial chickens - lay a clutch of 5–8 eggs once or twice a year and incubate them for 19–21 days (Del Hoyo et al., 2001). Once a hen has laid the last egg of a clutch and oestrogen levels decrease, the medullary bone is gradually lost and remodelling of the new cortical bone will be initiated (Whitehead, 2004). In modern layer strains producing 250–310 eggs in the first year after the onset of lay (Karcher and Mench, 2018), hens lay an egg on multiple consecutive days followed by only a few days without oviposition. As a result, modern commercial hens are not able to remodel and/or deposit a new cortical bone between two egg laying periods.

3 Identified bone health problems

Within commercial laying hen management, two major issues of bone health exist: keel bone damage (including deviations and fractures) and injuries during depopulations. Both are believed to relate to the underlying poor bone health and manifest in response to interactions with their environment and/or the handling required for the removal of hens at the end of lay (i.e. depopulation).

3.1 Keel bone damage: deviations and fractures

For laying hens, damaged keel bones, broadly classified as manifesting either fractures, deviations or a combination of the two, is one of the most important welfare problems facing the commercial laying hen industry as suggested by the UK's Farm Animal Welfare Committee (FAWC, 2010, 2013). The issue of keel bone damage has also been cited as a major problem by the EFSA-AHAW panel (European Food Safety Authority, 2015) and a North American-based consortium of welfare researchers (Lay et al., 2011).

Keel bone fractures can be defined as breaks in the bone that typically manifest as a callus around the fracture site but may also involve sharp, unnatural deviations or bending (Fig. 2; Casey-Trott et al., 2015). While fractures are considered to be a defect in the bone architecture, deviations are generally believed to result from the normal process of bone remodelling (Harlander-Matauschek et al., 2015). Deviations are defined as keels with an abnormally shaped structure that have not resulted from a fracture incident, but contain section(s) that vary from a theoretically perfect two-dimensional straight plane in either the transverse or sagittal planes (Fig. 3; Casey-Trott et al., 2015). Given that fractures are viewed as defects of the bone rather than a consequence of

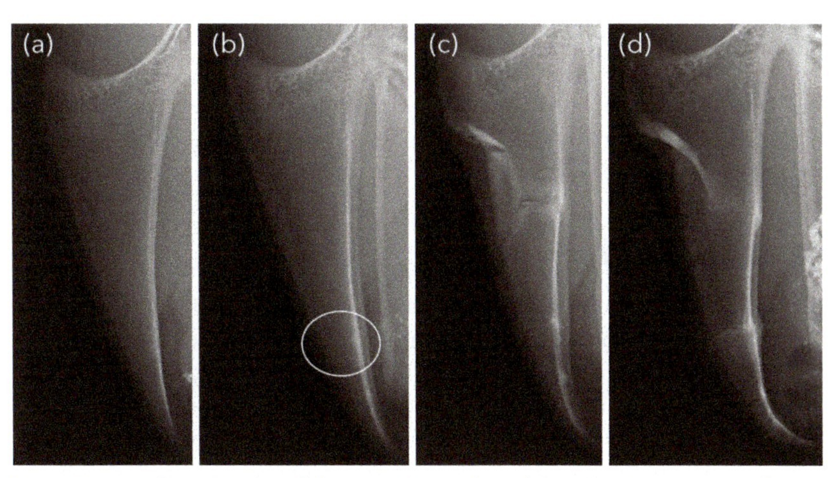

Figure 2 Latero-lateral X-rays of the same hen without fracture (a), with a minor fracture indicated by a circle (b), with multiple major fractures (c) and partially healed fractures (d). Source: C Rufener.

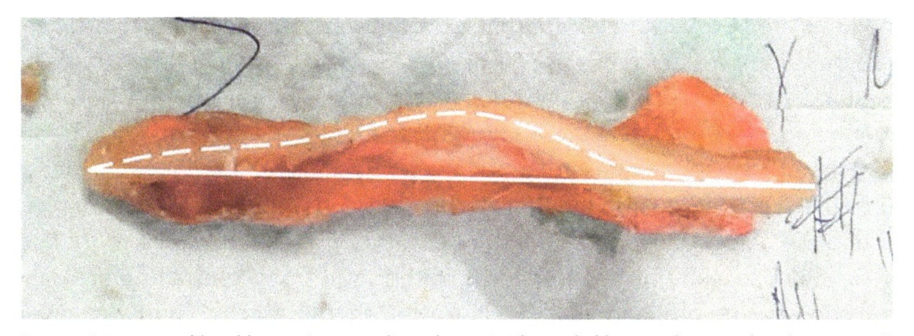

Figure 3 Deviated keel bone (ventrodorsal view). The solid line indicates the theoretically perfect two-dimensional straight transverse plane and the dashed line represents the deviated shape. Source: C. Rufener.

a normal physiological process (i.e. remodelling), fractures are more critical in terms of welfare and productivity. Deviations remain important to consider as a dramatically altered structure is likely to be weakened and more susceptible to fracture (Harlander-Matauschek et al., 2015).

Damage to the keel bone has been widely reported in multiple countries and levels are exceptionally high, typically in excess of 50% by the end of lay (Canada (Petrik et al., 2015); Belgium (Heerkens et al., 2013); Holland (Rodenburg et al., 2008); Switzerland (Käppeli et al., 2011); UK (Wilkins et al., 2011; Tarlton et al., 2013; Toscano et al., 2015)). In addition, the problem appears to occur in all types of housing systems (Wilkins et al., 2011), including conventional and enriched cages (Petrik et al., 2015) and organic systems (Bestman and Wagenaar, 2014), and across genetic lines (Käppeli et al., 2011;

Candelotto et al., 2017). The problem has also been observed in commercial broiler breeder flocks (Gebhardt-Henrich et al., 2017b). The underlying cause of keel bone fractures is generally considered to be high and sustained egg production leading to weak and brittle bones that are susceptible to fracture during collision, a theory discussed in more detail below.

3.2 Injuries during depopulation

At the end of a commercial laying period, all hens are removed from the barn in a systematic fashion to allow cleaning and preparation for the subsequent flock. The removal process, commonly referred to as depopulation, will occur typically between 60 and 90 weeks of age though sometimes up to 100 weeks of age depending on a host of factors including the flock's health, productivity and availability of the final processing system or location.

Despite the increased automation of tasks within the laying hen industry, the depopulation process remains largely a labour-intensive process requiring human workers to manually collect the hens within the barn. Once collected, the hens are normally crated for transport to the slaughter facility though can also be killed on-site for use as an energy source, for example, biogas (Afazeli et al., 2014). Beginning with the initial collection of hens in the barn and continuing to the arrival at the abattoir or death of hens otherwise, numerous occasions exist for injuries to occur including collisions with pen or cage furnishings or loading into the transport crate. Injuries are also likely to happen during the actual transport of the live hens, though the principal danger here is exposure to extreme temperatures, withdrawal from feed and water, and the consequent physiological stress (reviewed by: Mitchell and Kettlewell, 2009; Schwartzkopf-Genswein et al., 2012). Given that the collection of hens is considered the most stressful process as well as the source of injury during depopulation (Schwartzkopf-Genswein et al., 2012), the focus here is on the catching, carrying and crating of hens within the overall depopulation process. Similar to the concerns raised with keel bone fractures, the bones of the hens are believed to be relatively weak by this point making them susceptible to damage.

Focussing on hen injuries during collection, Gregory and Wilkins (1989) reported that an average of 24% of hens (ranging from 13% to 41% of farms surveyed) had a broken bone immediately following commercial depopulation which they attributed to the removal process. The seminal effort by Gregory and Wilkins (Gregory and Wilkins, 1989), as well as the majority of work in the area performed by others (Gregory et al., 1990; Kestin et al., 1992; Reed et al., 1993; Budgell and Silversides, 2004), focussed on laying hens from cage systems, although comparable levels of damage have been reported in aviaries (cage systems: 10.1%, Gregory et al., 1990; aviaries: 8%, Gerpe et al., 2017).

The removal process for aviary and cage systems has many parallel events and is likely to result in similar injuries as, in each case, the bird is typically forcefully pulled from the cage or their resting/perching position in the aviary. In this process, birds can collide with the edge of the cage as well as furnishings (e.g. perches, drinkers, etc.) either in the cage or that exist as part of the non-cage system.

Bones commonly found to be broken during removal from cages include those of the hips and legs: humerus, ischium and pubis (Gregory and Wilkins, 1989; Gregory et al., 1990). A similar category of bone injuries was also found to occur in aviaries, albeit with greater frequency in the keel, pubis and furcular with less to bones of the wings and legs (Gerpe et al., 2017). The reason for the variation in injury rates between housing systems is not clear though likely associated with multiple study-specific factors including genetic line (Budgell and Silversides, 2004; Candelotto et al., 2017), the increased load-bearing activity of hens in non-cage systems (Blokhuis et al., 2005) leading to increased bone strength (Leyendecker et al., 2005), as well as the differences in how hens are collected (i.e. pulled from within a cage in one direction vs. from multiple angles in aviary systems).

Once collected, birds are then bundled into groups of 3–5 hens and carried to a transport crate or other disposal device, for example, on-site gas chamber. While being carried, they are routinely held by a single leg which can result in twisting that can easily result in bone fracture or displacement. During the catching and carrying process, birds are frequently passed from a person performing the initial catching to a second person responsible for carrying and/or bundling, and possibly, a third person who actually loads the birds into the transport crate. Each passing of the birds between persons creates an additional opportunity for injury as the position of the birds is often rotated to accommodate the change in grasp. In support of the notion that improved handling during this stage of removal can lead to reduced injury, removal of birds by two hands versus one reduced injury from 14% to 5% (Gregory et al., 1993). Similarly, Gregory and Wilkins (1989) found that when birds were handled one bird at a time and brought to the transport crate, the rate of injury fell from 24% to 14%. During loading of hens, the keel and wings are especially susceptible to damage by collision with the sides of the transport crate (during placement into the crate) (Van Niekerk and Reuvekamp, 1994; Knowles and Wilkins, 1998). Mainly with broilers, a variety of commercial means have been pursued to reduce injury with moderate success including: legislation and better training of poultry handlers (Kettlewell and Mitchell, 1994) as well as technical improvements in how birds are collected and removed (Lacy and Czarick, 1998; Kristensen et al., 2001). Despite these efforts, laying hen removal remains a labour-intensive process with little or no advances in automation to reduce bird injuries.

4 Contributory factors to poor bone health

Skeletal injuries in laying hens are believed to result from compromised skeletal health though the underlying reasons of poor health are not fully understood. High and sustained egg production is considered the dominant factor, though recent evidence suggests the contribution of alternative factors as well.

4.1 Excessive calcium demand of egg production

The underlying cause that is most routinely associated with keel bone fractures during lay and injuries during depopulation is high and sustained egg production of modern commercial laying hens. To support the process of egg shell formation and the required calcium demand, bone resorption is used to supplement the exogenous (dietary) calcium required to form the eggshell. As mentioned above, bone resorption continues throughout the lay cycle with high rates of egg production. Over time, the production of eggs is believed to result in hens with a net loss of mineral in structural (cortical) bone leaving the bones weak and brittle (Whitehead and Fleming, 2000; Fleming et al., 2004; Bain et al., 2016). It is this loss of structural cortical bone that is thought to be responsible for the increased frequency of keel bone fracture as the hen struggles to support nearly continuous eggshell formation with diminishing endogenous mineral reserves. From the status of relative bone weakness, the hen is now susceptible to injury from collisions with barn furniture. Wilkins et al. (2011) suggested that the variation in fractures observed between systems was likely due to perches (and other system furnishings) in some systems resulting in more high-energy collisions, a view supported by the increased fracture occurrence and severity seen in systems with greater environmental complexity (i.e. more and higher items available for perching). Most studies have focussed on non-cage systems, though fractures do appear in caged systems (Wilkins et al., 2011) where high-energy collisions are also known to occur (Baker et al., 2020). Regarding multi-tier systems, reports have detailed collisions specifically within an aviary environment (Stratmann et al., 2015a), where hens had collisions as they sought positions at higher elevations, particularly during the dusk/dark transition (Stratmann et al., 2019). During this transition period, sitting hens occasionally get pushed from their positions by conspecifics or fail to land successfully, particularly where local bird density is very high.

The role of egg production as the underlying reason of keel bone fractures is supported directly by work using a synthetic gonadotropin antagonist to experimentally suspend lay resulting in reduced keel fractures (Eusemann et al., 2018b). Similarly, poultry bred for relatively high- and low-bone strength over seven generations found all males to be free of keel bone fractures whereas the proportion in hens ranged from 0% to 40% depending on the genetic line and generation (Fleming et al., 2004). Comparisons between high- and low-producing hybrids (Habig et al., 2017; Eusemann et al., 2018a), including heritage lines (Regmi et al., 2016a), also support the position of egg production as a cause of poor bone health. However, these studies are typically confounded by strain effects, for example, differential behavioural activity between genetic lines.

Although egg production is likely the major contributing factor to keel bone fractures, the problem appears to pre-date modern intensive production programmes with the earliest published observations made by Darwin (1868) reporting that 88% of chickens (both males and females) across 12 commercial breeds had a deformed sternum. Later reports by Warren (1937) and Buckner et al. (1949), though still preceding modern selection programmes, highlighted keel bone damage with the suggestion that the cause was related to genetics and the availability of perches and vitamin D. More recently, within the last 3–5 years, methodological advances have allowed for a more reliable and detailed assessment of keel damage with specific techniques including: improved palpation training (Casey-Trott et al., 2015; Gebhardt-Henrich et al., 2019); routine use of radiographic imaging (Baur, 2017; Eusemann et al., 2018a; Rufener et al., 2018a,b, 2019; Gebhardt-Henrich et al., 2019) and computed tomography (Candelotto et al., n.d.; Regmi et al., 2016b; Chargo et al., 2018a,b).

Improved assessment, in combination with the use of an ex vivo model that used an impact testing apparatus to experimentally create fractures in deceased birds (Toscano et al., 2013, 2018), has suggested alternative or additional influencing factors that may be contributing to the development of keel bone damage. The additional factors may operate in addition to or in synergism with underlying poor bone health in response to high and sustained egg production (Toscano et al., 2020). In support of the role of other factors beyond egg production causing keel bone fractures, the use of an impact-testing protocol identified a post-peak egg production phase where, beyond 49 weeks of age, susceptibility to new fractures appeared to decrease. The finding was surprising given that egg production remains relatively high (> 85% hen daily percentage) and that hens are not believed to produce cortical bone while they remain in lay (Hudson et al., 1993). Nonetheless, the findings parallel to age-related (i.e., longitudinal) keel bone fracture frequencies reported independently in commercial flocks where rates of fracture often stabilize or decrease in the same period (Richards et al., 2012; Petrik et al., 2015; Stratmann et al., 2015a; Toscano et al., 2015). The stabilization or decrease in new fractures could be attributed to altered behaviour or improved navigation abilities, though the impact testing data, in combination with similar patterns in commercial flocks, suggests some qualitative change within the animal such as bone maturation and/or age-related changes in strength (Rath et al., 2000). In other words, if egg production was directly related to bone health and occurrence of fractures, the rates of fracture would continue increasing beyond 49 weeks of age, which does not seem to be the case. Rather, the rate of new fractures appears to level-off or possibly even decrease in both commercial and experimental settings. In further support of the unclear role between egg production as the underlying reason of keel bone fractures, hens showing a decreased susceptibility for fracture had greater egg shell width and strength (Candelotto et al., 2017), a finding in conflict with the expectation that egg production and bone health are in direct competition.

4.2 Alternative contributing factors to keel bone fractures

Beyond intensive egg production, the length of bone development in the period preceding puberty, that is, the age at first egg, has been suggested to influence bone quality (Wolc et al., 2010), which in turn is influenced by growth rate, photoperiod and body weight (Dunn and Sharp, 1992). As the keel is understood to ossify late compared to other parts of the skeleton (Buckner et al., 1948) and remains cartilaginous even at puberty (Casey-Trott et al., 2017b), a premature transition to the lay period before completion of bone development (Rath et al., 2000) could prevent the keel from reaching peak strength in terms of its full genetic potential. Given that the keel ossifies cranial to caudal leaving the tip to be the final portion to mature (Buckner et al., 1948), which does not finally ossify until an estimated 35–40 weeks of age, the delay in maturity could explain reports where the tip (i.e. most caudal aspect) of the keel also manifests the highest rates of damage (Riber and Hinrichsen, 2016; Baur, 2017; Baur et al., 2020). A comparison of traditional and commercial genetic lines (Hocking et al., 2003) found the former come into lay an average of 4 weeks later while also possessing improved bone health, though results were confounded by breed differences such as differences in body mass. By manipulating photoperiod with a small but definitive delay of four days, Silversides et al. (2006) was able to induce positive effects on medullary and cortical density and area of the radius and humerus. Hester et al. (Hester et al., 2011) found no effect of pullet lighting on end of lay bone mineral content suggesting the effects may be short-lived, though they were unable to document whether there was a latency to enter lay between treatments. Interestingly, Gebhardt-Henrich and Fröhlich (2015) examined natural variation in fracture occurrence and found that hens without evidence of fracture at end of lay came into lay 17 days later than hens with fractures. In summary, there is evidence to suggest that premature occurrence of lay could impinge on a hen's ability to develop a mature skeletal structure, which could lead to increased susceptibility for keel bone fractures.

Other possibilities to explain keel bone fractures include the existence of underlying pathological conditions that may resemble an injury that would result from a collision. Most work to date assessing fractures has used rather coarse techniques (i.e. palpation and direct examination of excised keels (Casey-Trott et al., 2015)), with the presence of callus material considered as positive identification of fracture. However, a callus-like expansion of bone can also result from other processes such as periosteal irritations, congenital malformations, periostitis, osteomyelitis and bone tumours (Craig et al., 2016). In this scenario, while more severe or traumatic fractures may result from collisions, there is also a possibility for pathological fractures or a predisposing disease resulting in a weakened skeletal architecture (in contrast to the skeleton being weakened by high egg production). These predisposed bones would be expected to break more easily, even with routine daily forces that would generally be considered innocuous. In contrast, traumatic fractures occur only with an increased/massive force acting on normally structured bones (Craig et al., 2016), though this work has yet to be studied in terms of fractures to the keel. Pathological conditions could include diseases of a genetic (e.g. osteogenesis imperfecta, or glass bone disease), metabolic (e.g. osteoporosis) or nutritional (e.g. osteodystrophia fibrosa, rickets) origin, or bone tumours. Each of these pathological states could result from factors independent of egg production while yielding conditions that make bone susceptible to fracture. It is even possible that, as a result of the intensive selection for egg production, linked traits that relate to these pathological conditions could have been inadvertently selected in the process via a form of genetic hitchhiking.

5 Influence of poor bone health on productivity and welfare

Keel bone fractures appear to have detrimental effects on hens' productivity and welfare (Riber et al., 2018). In terms of productivity, fracture repair was suggested to involve a diversion of resources where minerals (e.g. calcium) and energy, normally directed towards the reproductive process of egg production, now must be reallocated to the process of healing bone (Thiruvenkadan et al., 2010). In support of this prediction, hens with increasing severity of keel fracture as defined by a radiographic protocol had decreased egg production that became visible from approximately 50 weeks of age and increased in magnitude with age (Rufener et al., 2018a). The greatest modelled difference between the least and most severe manifestations of damage by 61 weeks of age was 16.2%, meaning that 61-week-old hens with the most severe keel bone fractures would be expected to lay 16.2% fewer eggs than 61-week-old hens without fractures. Similar findings have been reported by others who identified an associated increase with feed and water consumption (Nasr et al., 2012a), although this report did not control for individual variation. Interestingly, assessments done at the flock level were unable to identify a relationship between flock production and the percentage of hens with keel fractures (Heerkens et al., 2013). The reason for the lack of a consistent relationship between egg production and fractures is not clear, where the ambiguity is likely worsened by: difficulties in assessment by palpation (Casey-Trott et al., 2015), high-variable rates of fracture incidence at the flock level, as well as the high, individual-level variation of egg production combined with the suspected/predicted relatively small effects of keel fractures on egg production (Candelotto et al., 2017; Rufener et al., 2018a). Farm-level management factors that are difficult to quantify, such as feed quality or efforts to calm hens, are also likely to contribute to variation and handicap efforts at linking egg production and keel bone fractures.

Parallel to the correlations between the presence of fractures and productivity, changes in mobility can also be indicative of pain and compromised welfare (Prunier et al., 2013). Behavioural indicators of pain are normally characterized by reactions, which serve to terminate the causative stimulus or reduce its effects (Mellor et al., 2000). Specifically in regard to keel bone fracture, Nasr et al. (2012b, 2015), using birds trained to descend from a 50 cm high perch for a food reward, found birds with fractures manifested an increased latency to descend in comparison to those without, a response which was attenuated with analgesics suggesting the involvement of pain pathways. In this scenario, the latency to descend from the perch was reasoned to involve the bird's expectation of pain that would result from musculature attached to the keel generating force during motion as well as the keel making contact with the floor during descent. The observed latency may reflect an interplay of conflicting motivations - the first, where the hen's eagerness to consume the reward competes with the second, the desire to avoid the likely pain that would result. If this mechanism is occurring, it reflects an interference of normal locomotive behaviour and raises the possibility of compromised welfare. As an extension of Nasr's training paradigm, Rufener et al. (2019) demonstrated altered movement patterns between defined zones (i.e. aviary tiers and an external wintergarden) with increasing severity of keel bone fracture. Hens with damaged keels spent more time in the uppermost tiers containing resources (feed, water, nest box) for which demand is generally inelastic. In contrast, these hens also made fewer visits to the lower levels containing resources (e.g. litter for dust bathing) for which demand is more elastic, suggesting definitive behavioural changes in response to fracture. Similar responses in terms of nest box usage (Gebhardt-Henrich and Fröhlich, 2015) and accessing the outdoor (i.e. in this case, free range pasture) environment (Richards et al., 2012) have been reported. Movement within cage systems, although associated with less severe keel damage, results in altered behaviour (Casey-Trott and Widowski, 2016). Taken together, these results support the position that hens manifesting keel fractures within cage and cage-free systems are adopting their behaviour to avoid pain-associated with movement.

Interestingly, specific aspects of fractures such as a fracture gap relate to decreased nest and perch access (Rentsch et al., 2019), suggesting particular types or features of fractures may be more critical than others. Breaks (or gaps) in the keel, and by extension the periosteum, which have the highest density of sensory and sympathetic nerves (Mach et al., 2002; Nencini and Ivanusic, 2016), would result in the greatest sensation of pain (Freeman et al., 2008). More work defining the specific features of keel fractures and their associated behavioural responses, ideally in concert with analgesic manipulation, is needed to determine how different types of fractures relate to pain and consequent welfare.

6 Strategies for improving bone health

The problem of keel bone damage is a complex and multifactorial issue, thus solutions will need to consider these various causes in detail. Furthermore, it is likely that no single strategy will suffice, but rather a combination of strategies - genetic, nutritional and management - tailored to the specific housing and production environment will be needed.

6.1 *Effective breeding strategies*

In considering potential tools for genetic selection against keel fractures, it needs to be determined which traits should be selected where possibilities may include: stronger bones, improved physical ability, increased mineral absorption, rates of bone remodelling and increased docility. Strategies to link traits with keel fractures have suggested that keel and humerus bone mineral density and strength are associated with reduced keel fracture susceptibility (Candelotto et al., 2017, n.d.) as well selection of specific QTLs (Dunn et al., 2007; Podisi et al., 2012). Alternatively, as susceptibility to bone fractures are a relatively complex genetic trait, genomic selection is also an option (Fulton, 2012). Genomic selection involves monitoring the incidence of fracture in a large population of laying hens and comparing the genomic information of the two sub-populations. One of the strengths of this approach is that it does not target a single factor, for example, bone strength, but focusses on the presence or absence of damage. As part of the process, the mechanism associated with the trait of interest is selected. A challenge with the genomic approach is that it requires a large sampling population (>5000) of which a clear phenotype is needed. Given recent emergence of the complexity of damage that actually exists (Baur et al., 2020), selection of a clear fracture-free phenotype may be problematic. However, if successful and a genomic phenotype is identified, no further phenotypic measurements are required for the selection programme (Eggen, 2012). Although breeding for reduced keel bone damage is attractive, efforts need to be made to ensure that only the traits of interest are selected with minimal loss of desired traits, for example, shell quality.

6.2 Nutritional strategies

Efforts to improve bone strength have principally involved better calcium delivery for egg production or slowing the process of de-mineralization. One example that combined these strategies was use of omega-3 fatty acids, which can modulate the processes that drive bone de-mineralization by decreasing the activity of osteocytes, the cells responsible for bone resorption (Liu et al., 2003; Sun et al., 2003; Griel et al., 2007; Shen et al., 2007). Omega-3 fatty acids also enhance the formation of bone by osteoblasts (Watkins et al., 2003). Intervention with omega-3-enhanced diets results in reduced fracture incidence (Tarlton et al., 2013; Toscano et al., 2015), but is also known to benefit human health (Calder, 2006, 2009) leading to benefits for human consumers as well. There is also evidence from various animal studies, however, that omega-3 fatty acids can have detrimental effects on health, possibly through interactions with immune function (Anderson and Fritsche, 2002) or production of damaging free radicals (Aruoma, 1998). The detrimental effects may also be relevant for laying hens (Toscano et al., 2012, 2015).

A second strategy is use of larger calcium particles which take longer to break down and thus maintain elevated dietary calcium throughout a 24-h period. Provision of limestone particulate, oyster shell or other forms of calcium with a relatively large particle size and thus requiring a longer duration for digestion, can maintain egg-shell quality and extend the lay period up to 100 weeks of age (Thiele, 2015; Pottgüter, 2016). Egg shell formation occurs in the evening when the hen is sleeping and is associated with a nine-fold increase in resorption (Van De Velde et al., 1985) to provide greater calcium to supplement dietary sources as needed. Larger dietary calcium particles can thus increase the amount of calcium available during this critical night time period. Although published work on this strategy has focussed on egg shell quality, the reduced need for endogenous calcium, for example, from bone, would likely extend to the keel and thus aid in maintaining skeletal integrity. In other words, as greater amounts of calcium would be available in the blood from the diet, less would be required from the bone. Beyond the form of calcium, feeding the daily requirement of calcium towards the end of the day and/or the last feeding has also been suggested as a strategy to increase calcium concentrations at night and reduce the need for bone resorption (Thiele, 2015).

6.3 Rearing hens for better cognitive and musculoskeletal development

The rearing phase is crucial to prepare pullets for their life in the layer house, particularly in the case of more complex (e.g. furnished cages, open barn) and multi-tier housing. In terms of bone health, the environment a pullet is exposed to during her early life affects the musculoskeletal development as well as spatial cognition and behavioural experience with three-dimensional space, which in turn relates to both susceptibility to and the actual occurrence of fractures. In other words, the rearing environment affects the musculoskeletal strength, and thus the susceptibility to fractures when exposed to trauma, for example, due to a fall or collision. The likelihood for falls or collisions, on the other hand, relates to the hen's ability to move through vertical space which is modulated by her behavioural and cognitive development during early life.

6.3.1 Musculoskeletal development

Physical exercise leads to higher load-bearing activity on the skeleton which in turn enhances bone strength (Lanyon, 1992). It is well established that housing systems offering opportunities for physical activity (e.g. walking, running, flying and jumping) are associated with improved bone health compared to cage systems, which allow comparatively limited exercise and bone loading (reviewed by Webster, 2004).

The link between the physical activity and bone strength has been well established in adult hens. In humans, the effect of physical exercise on bone strength is strongest before the onset of sexual maturity (Bass, 2000). Correspondingly, the rearing phase (i.e. from hatch to approximately 17 weeks of age) is believed to be a particularly crucial phase regarding bone development in laying hens - especially because the formation of structural bone ceases once a hen enters lay (Hudson et al., 1993). Compared to pullets reared in conventional cages, pullets reared in cages containing perches or in aviaries had improved bone health as indicated by a variety of responses including: increased long bone width and keel length (Casey-Trott et al., 2017b), increased bone mineral density or content (Enneking et al., 2012; Casey-Trott et al., 2017b), and higher cortical bone density, ash content, stiffness and strength (Regmi et al., 2015).

Importantly, increased bone strength seen at the end of rearing seems to persist throughout lay even if hens were transferred to cages after rearing, that is, with minimal bone loading during the laying phase (Casey-Trott et al., 2017c). Hester et al. (2013) also found greater bone mineral density at the end of lay in caged hens that had access to perches only during the rearing phase. Michel and Huonnic (2003) reared pullets in aviaries or floor pens with perches before moving them into aviaries and found higher breaking strength of tibia and humerus in aviary-reared hens at the end of lay. Increased bone strength in aviary-reared, but cage-housed hens further resulted in a lower keel bone fracture prevalence at the end of lay (Casey-Trott et al., 2017a). However, as a suggestion of the limits for enriched rearing, increased mineralization of the keel seen as a result of perch access during rearing did not prevent higher keel bone fracture incidence at the end of lay if hens were given access to perches during the laying phase (Hester et al., 2013).

Besides promoting skeletal development, physical exercise increases muscle mass. Casey-Trott et al. (2017c) found heavier breast and wing muscles in aviary-reared pullets compared to pullets reared in conventional cages, though cage-reared pullets had heavier leg muscles. As with bone health, this effect seems to last throughout the laying phase as Hester et al. (2013) showed increased muscle deposition at the end of lay when hens were given access to perches as pullets. In an experimental study using an ex vivo protocol to model bone fractures resulting from collisions (described previously in this chapter), Toscano et al. (2018) found that breast muscle mass was related to a lower risk to obtain major and severe experimental fractures. The benefits of breast muscling could relate to breast muscles serving as cushion if a keel bone is exposed to trauma, for instance, when a hen collides with a perch. As total body mass is associated with keel bone fracture prevalence and severity (with heavier birds having more severe damage at the keel; Gebhardt-Henrich et al., 2017a) the nature of the relationship is not clear. Breast muscle mass could have a cushioning effect resulting in a reduced impact on the bone, or increased muscle mass could link with higher load bearing on the keel and therefore a stronger bone being less susceptible to break.

Overall, multiple studies have shown that the properties of the rearing system – especially the provision of perches and access to vertical space – affect the physical properties of the skeleton and lead to a lower susceptibility for fractures when exposed to trauma. In addition, housing systems promoting exercise and therefore muscle deposition, especially breast muscles around the keel, might affect skeletal integrity positively due to a reduced risk for injury at the keel. Providing access to resources allowing exercise such as flying and perching during the rearing phase is therefore a powerful strategy to enhance bone health with lasting effects throughout the laying period. As there seems to be only a limited association between bone traits and fracture prevalence (Gebhardt-Henrich et al., 2017b), non-skeletal factors such as behaviour and spatial abilities are believed to strongly influence the risk for injuries at the keel bone (Toscano et al., 2018).

6.3.2 Spatial abilities: locomotor skills, perception and cognition

In aviaries where resources (e.g. food, water, nest boxes, perches and litter) are distributed over several stacked tiers up to 3.5 m high, a hen must be able to move within her environment to access these resources. Failure to coordinate movement successfully and consequent falls and collisions within these complex housing systems are likely to result in trauma resulting in keel bone fractures (Sandilands et al., 2009; Wilkins et al., 2011; Harlander-Matauschek et al., 2015; Stratmann et al., 2015b). Although the development of spatial abilities is important for all housing systems, the complexities of aviary systems with their multiple vertical tiers combined with the dangers of poor movement coordination and locomotor skill are the focus of this section. Complexity of the housing environment is an abstract term, but here it is intended to convey the relative amount of resources within housing across multiple areas that require memory and skill to move between (horizontally and vertically), typically with a variety of movement types (e.g. jumping, walking, and flying).

To successfully move within three-dimensional space and avoid falls and collisions, pullets need to have appropriate: (visual) perception, cognitive abilities involving processing of visual and spatial information, and locomotor skills and physical strength. Perception involves visual acuity or depth perception, whereas the cognitive abilities refer to processes such as memory, the discrimination of local and global cues, route planning or knowledge of location. Locomotor skills and physical strength are closely related to the musculoskeletal development described earlier, but require further training of specific locomotor patterns. For instance, while a certain range of locomotor skills and physical strength is needed for a pullet to run horizontally in a floor housing system, a different range must be developed for the jumping and flying to a perch required in an aviary system.

The behavioural component of successfully moving within a housing system is intertwined with cognition and perception. As an example, independent of a hen's strength, a hen needs proper visual acuity and depth perception to assess flying distance as well as spatial memory and knowledge of location to orient herself. In turn, a hen needs to gain experience with specific locomotor patterns to develop the corresponding cognitive abilities. For instance, a hen has to access an elevated platform first to develop knowledge of location corresponding to her position in vertical space. The following sections describe the development of perception, cognition and locomotor skills and give an overview of how housing complexity during rearing could affect these aspects of spatial abilities.

To successfully move within structures of a housing system, a hen needs to be able to perceive her environment properly. Perception involves visual acuity, depth perception, spectral and flicker sensitivity, and accommodation (the ability to change focus). Compared to single-tier or cage-systems, living in an aviary system requires greater depth perception and contrast sensitivity, as the ability to assess distance and angles is crucial to move between vertical elements successfully. Exposure to light plays an important role regarding normal eye development (Prescott et al., 2003), but it remains unclear how other environmental factors affect perception. The visual system in chicks seems to be fully developed by two days of age (Over and Moore, 1981). At this age, chicks can avoid obstacles, follow moving objects and peck at objects. As an indication of depth perception at an early age, newly hatched as well as two-day-old chicks discriminate between shallow and deep surfaces on a visual cliff (Walk and Gibson, 1961; Green et al., 1993). The early environment seems to affect depth perception to some degree (Tallarico and Farrell, 1964), but the underlying mechanisms have not been investigated.

Visual and spatial cognition also develops early in life. Aspects of cognitive abilities such as the interpretation of object permanence, amodal completion, use of spatial cues, memory or knowledge of location have been discussed elsewhere in this chapter. In brief, spatial cognition develops with age, and lateralization plays an important role. A left hemisphere bias develops in ovo and is dominant throughout the first day of life. As a result, a chick's visuospatial analysis initially focusses on local features (e.g. to approach objects or discriminate food items; Vallortigara and Rogers, 2005). Around 11 days of age, chicks would begin to move out of sight of the hen, which coincides with a shift of bias towards the right hemisphere (Freire and Cheng, 2004). Accordingly, chicks start to recognize global cues and relational properties of spatial layouts after 11 days of age (Vallortigara et al., 1997).

The degree of spatial information available in the environment affects the cognitive development of chicks. For instance, rearing chicks with visual barriers resulted in better/faster performance in spatial memory tests (visual displacement and detour tests) when tested at 11 days of age (Freire and Cheng, 2004). In more applied settings, pullets reared in aviaries had a better working memory than pullets reared in cages (Tahamtani et al., 2015), and hens not having access to perches within the first eight weeks of their life had longer latency to find a food reward via platforms with increasing difficulty of the task (Gunnarsson et al., 2000). As the Gunnarsson study did not find any differences in locomotor skills for the easiest task (i.e. reaching a platform 40 cm above ground) and because all hens used vertical space via perches in their home pens, the authors concluded that hens with early access to perches had improved spatial cognition rather than just better locomotor skills (compared to hens that had access to perches only after 8 weeks of age). Further, chicks reared with access to perches and elevated structures were faster than floor-reared chicks in a test requiring chicks to navigate a detour in order to reach conspecifics (Norman et al., 2019), a task where differences in physical strength are unlikely to play a major role.

Locomotor skills develop relatively early during the rearing phase and map onto cognitive changes related to the shift in hemisphere bias. Accordingly, if given the opportunity to access three-dimensional space, individual chicks accessed perches of 10–40 cm height as early as 8 days of age (Gunnarsson et al., 2000; Heikkilä et al., 2006; Wichman et al., 2007), though most studies showed that the majority of chicks used elevated structures such as perches and platforms around 14 days of age (Heikkilä et al., 2006; Wichman et al., 2007; Kozak et al., 2016; Habinski et al., 2017).

If pullets or hens have not gained experience accessing vertical structures such as perches or raised tiers at an early age, they struggle to utilize the full area of complex housing systems as adult hens. Hens reared in aviaries are distributed on all tiers in experimental pens, whereas hens reared in cages spent most time on the floor (Brantsæter et al., 2016). Along those lines, Colson et al. (2008) compared floor- and aviary-reared birds that were housed in an aviary system during lay and reported that hens reared on the floor used upper levels less, showed reduced accuracy in long flights and jumps, and preferred to stay on litter and lower levels. Similarly, hens reared on the floor used the highest tier of the aviary in the layer house less frequently than aviary-reared hens (Michel and Huonnic, 2003).

Despite the evidence showing that space use in the adult housing system is affected by the complexity experienced during rearing, it is difficult to disentangle the underlying mechanisms. In other words, differences in space use in response to the rearing system are probably not only related to the locomotor skills and physical strength of a hen, but more likely the result of interactions between locomotor skills, cognitive development and possibly perception.

In addition to the effects of general housing complexity on spatial abilities, the lack of experience with specific housing-system elements (e.g., ramps) affect the ability to negotiate transitions in a novel environment even if pullets had access to vertical space during rearing. Hence, it is important that the rear and lay environments are considered together. Norman et al. (2018) reared pullets with perches and platforms, but half of the birds were additionally provided with ramps (Fig. 4) to facilitate platform access. During testing at 12–14 weeks of age, pullets that had access to ramps were more successful in reaching food rewards on a platform connected with a ramp and took less time to complete these upward transitions. In addition, ramp-reared pullets showed double the number of transitions between the litter and platforms during group-housing observations immediately after being moved to a novel environment. The study by Norman et al. indicates that both general access to vertical space as well as experience with specific structural elements influences the locomotor skills within three-dimensional space.

In summary, spatial abilities obtained from interactions with the rearing system are highly relevant for laying hens being housed in complex environments. Environmental complexity and provision of structures enabling the access to vertical space promote necessary cognitive development and allow training of locomotor skills, which are critical in aiding pullets as they are moved from the rearing facility to the laying environment. The transition from

Figure 4 Pullet using a ramp within a commercial multi-tier housing system. Source: A. K. Rentsch.

the pullet to the layer house has a myriad of challenges as hens must - again - learn to navigate within their novel environment.

It is the general consensus that differences between the rearing and the laying environment should be reduced (reviewed by Janczak and Riber, 2015), and meeting a pullet's cognitive and behavioural skills obtained during the rearing phase is crucial to maintain good bone health. Ideally, a hen placed in a novel environment will have had previous experience with all elements of the housing system (perches including specific materials and shapes, ramps, platforms), and had the opportunity to learn how to transit between tiers to reach all resources successfully. To prepare pullets optimally for their novel environment as a laying hen, it is therefore important to not only rear for a specific laying environment, but to enhance the laying environment in order to facilitate adaptability to the laying house.

6.4 Adjustments to layer housing

Even though the design and management of the rearing environment can contribute to the cognitive, behavioural and musculoskeletal development of pullets, it is unlikely that stronger bones and enhanced navigation abilities obtained during rearing can reduce keel bone fracture frequency and severity on their own. Adaptations of the laying environment are therefore - in addition to the provision of an optimal rearing environment - necessary to improve the skeletal integrity and reduce keel bone fracture prevalence in adult laying hens maintained under commercial conditions.

6.4.1 Complexity of modern housing systems and the role of perches

Modern housing systems such as aviaries can be up to 3.5 m high with several stacked tiers providing different resources via perches or platforms, which require the hens to jump or fly in order to reach the next tier and the associated resource. While aviaries benefit the hen by providing specific resources for highly motivated, species-specific activities (Weeks and Nicol, 2006), their increasing adoption by producers is also related to the economic benefit of being able to place more hens within a given area (Fröhlich and Oester, 2001; Aerni et al., 2005).

Domestic laying hens are descendants from the red junglefowl (Fumihito et al., 1994), a generally ground-dwelling species, and are therefore not particularly well suited to move within these complex environments. More so, increased body weight relative to wing area as a consequence of focussed genetic breeding has most probably led to a hen that lacks the stability of its ancestor and thus may have greater problem moving within aviary systems (Sandilands et al., 2009). Thus, it is not surprising that hens might have difficulty descending within an aviary system. In commercial aviaries, angles and distances between aviary elements often exceed the navigation abilities of laying hens which might explain why Campbell et al. (2015) observed up to 21% of flights in an aviary resulting in failed landings. Failed landings and/or falls can result in crashes and collisions which are assumed to be one reason for the high prevalence of keel bone fractures in laying hens (Campbell et al., 2015; Stratmann et al., 2015a,b). Similarly, a positive relationship was seen between the inter-row distance of aviaries and rates of keel bone fractures, which the authors reasoned resulted from failed attempts to jump between rows (Heerkens et al., 2015). Perches are especially seen as a hazardous element in aviary systems due to their exposure increasing the risk for high impacts after uncontrolled falls and collisions (reviewed by Sandilands et al., 2009).

Even though perches are suspected to be directly related to keel bone fracture occurrence, they are an important element in a layer housing system which is a paradox highlighted previously (Sandilands et al., 2009). Perches not only offer the opportunity to fulfil the behavioural need for roosting in elevated locations, but also serve as a means to access the stacked tiers of an aviary and thus, allow for movement within the system (reviewed by Struelens and Tuyttens, 2009). Perch material and shape are known to mitigate the problem of perches being a hazardous element. Perch shape affects keel bone health, with circular perches causing more damage to the keel than rectangular perches (Tauson and Abrahamsson, 1994). In the same vein, Pickel et al. (2011) measured the peak force experienced at the keel in hens perching on different perch types. Round and oval perches resulted in the highest force, followed by squared and an experimental, cushioned prototype perch. The authors suggested that increased contact area, thus, reduces localized pressure on the keel as well as a soft surface may reduce keel bone problems.

Within more applied settings, use of soft perches and reduced keel bone damage is supported by Stratmann et al. (2015b) who found a reduction in keel bone fracture prevalence and hazardous landings (Scholz et al., 2014). Perch design may also increase landing safety as Pickel et al. (2010) reported fewer balancing movements with increasing perch diameter (27, 34, 45 mm) on rubber compared to wood and steel perches. Perch material, as well as cleanliness, is likely to influence overall ability to move as the latency, pre-jumping behaviour, slips, failures to jump and crashes were influenced by jumping from wooden, metal and PVC perches (Scott and MacAngus, 2004). It is suggested to install non-slippery perches with a large contact area, possibly made from a soft material, to reduce keel bone fracture prevalence. To date, the development of such perches and the application to practical farming have proven difficult due to problems with hygiene and pest management (e.g. infestation of soft surfaces by mites). Thus, not only the shape and material of perches, but also their placement must be considered in order to facilitate bird movement within a given housing system.

6.4.2 Facilitating bird movement

Depending on distance and angle between take-off and landing sites in a commercial non-cage system, laying hens may have difficulty moving between resources and tiers successfully. Under experimental conditions, angles steeper than 60° for upward movements (Scott et al., 1997) and steeper than 30° for downward movements (Lambe et al., 1997; Scholz et al., 2014; Scott et al., 1997) have been shown to result in falls at landing, whereas a horizontal distance of more than 50 cm seems to increase the difficulty to navigate (Scott and Parker, 1994). Downward paths appear to be more difficult for laying hens than upward paths (Moinard et al., 2004; Scott et al., 1997), and the presence of conspecifics or obstacles at the landing site resulted in an increased risk for falls (Moinard et al., 2005). Lighting conditions only affect landing accuracy and the latency to jump when the intensity is very low (i.e. below 2 lux; Moinard et al., 2004; Taylor et al., 2003). Even if navigated successfully (i.e. without a fall at landing), increased angles and distances as well as downward compared to upward movements result in higher forces experienced at the keel (Rufener et al., 2020). Accordingly, the European Food Safety Authority (EFSA) reviewed the welfare aspects of perch use for laying hens extensively and stated that the risk of injury increases when hens have to jump a distance of more than 80 cm vertically, horizontally or diagonally, or jump an angle steeper than 45 cm (EFSA Panel on Animal Health and Animal Welfare, 2015). It is therefore recommended to install perches in a way to allow transitions of a short distance and flat angle.

However, angles and distance in commercial aviaries are often steeper and longer than recommended, and rearranging perches in a given housing system can be difficult as they often serve as a supporting structure. Several studies have proposed alternative solutions to facilitate hen movement within aviaries. Stratmann et al. (2015a) modified a commercial aviary setting by adding perches, platforms or ramps to facilitate movements between the tiers of the system, aiming to reduce the risk for falls and collisions. The authors found fewer falls and collisions as well as a lower keel bone fracture prevalence in hens provided with ramps and suggested that the continuous path between the tiers supports safer and more appropriate behaviours such as walking and running instead of flying and jumping. Accordingly, Heerkens et al. (2016) found similar effects in an experimental setting, where ramps were used to connect stacked tiers and perches at different heights resulting in fewer keel bone fractures. As different strains show considerable differences in the spatial distribution in vertical space (Ali et al., 2016; Kozak et al., 2016), it is important to work on solutions facilitating movements irrespective of strain.

Overall, the installation of ramps (Fig. 4) is a relatively inexpensive and straightforward adaptation of most non-cage systems which could improve navigation and consequently reduce fracture prevalence. The opportunity to walk instead of fly or jump might be especially beneficial for hens already having keel bone fractures as the wing muscles are attached to the keel (Sisson et al., 1975) and because fractures have been associated with limited bird movement in aviary systems (Rentsch et al., 2019; Rufener et al., 2019). Flying, wing-assisted running, jumping and other activities involving wing muscles are likely to apply higher forces to the keel presumably resulting in more pain if a fracture is present. Hens can climb ramps without using their wings up to an angle of 40° (LeBlanc, 2016), thus the installation of ramps with an angle up to 40° could allow hens having keel bone fractures to access upper levels and associated resources with less pain. Accordingly, Rentsch et al. (2019) found that hens with fresh fractures were less likely to show vertical locomotion than hens with absent or healed fractures, but demonstrated that hens used ramps irrespective of keel bone health. As hens can walk on grid-type ramps but need to jump from rung to rung on latter-type ramps (Pettersson et al., 2017; Norman et al., 2018), the installation of grid ramps at an angle of <40° can facilitate transitions between tiers and reduce keel bone fracture risk.

6.5 Improving hen handling during depopulation

To improve welfare during hen removal, it is key to reduce the amount and duration of manual handling (Scott, 1993). For short periods or with small flocks (e.g. less than 200 birds), catchers are typically able to perform the task with minimal injury to the birds and themselves. However, once the duration of removal is extended and fatigue does become a factor, it becomes difficult for even the most devoted of caretakers to maintain the appropriate concentration and physicality for such a task resulting in animal suffering and worker hazard (Kettlewell and Mitchell, 1994). For instance, as an example of attempts to reduce the human factor of fatigue in hen removal, Kristensen et al. (2001) reported the development of a modular system on wheels that brought the transport crate directly to the birds and minimized the carrying process, reducing the handling time for 450 hens from 65 min to 4.5 min. Despite the increased speed of removal, the technique resulted in a statistically similar (though numerically less) frequency of bone breakage. As only two farms were assessed and the collectors were aware of the research objectives, this may have impacted observed results. Along the same vein, in the last 60 years, various formats of collection techniques have been developed to reduce the human component including: a recessed conveyor belt that moved the birds directly to the truck, mats that could be laid down prior to collection that would be lifted out of the house with birds contained and large-tined forks that could scoop birds off the floor (Lacy and Czarick, 1998). While each of these mechanical efforts have held some degree of promise, none has been widely adopted by the industry due to a variety of reasons including: excessive capital costs, poor integration into existing systems, discomfort with the new technology, among others (Scott, 1993). Nonetheless, improving the quality of handling by reducing the fatigue of the collectors is key to improving the welfare of removed hens (Gregory and Wilkins, 1989; Kettlewell and Mitchell, 1994).

7 Future trends in research

Given the existing knowledge that has been generated, the scientific community is poised to make rapid advancements in understanding and resolving the problem of keel bone damage. One key step has been advances in detection, which can be done rapidly and within commercially relevant settings (e.g. by standardizing palpation and through radiographic images). While not providing direct benefit, improved assessment will allow for accuracy to better evaluate intervention strategies of nutritional, genetic and management natures. Another key area will be greater focus on developmental processes during rearing, particularly with increased use of cage-free systems and the appreciation of how birds must develop to use and benefit from the contained resources.

8 Where to look for further information

8.1 Further reading

An additional resource focussing on keel bone damage is the chapter 'Skeletal problems in contemporary commercial laying hens', Advances in Poultry Welfare Science, 2018. General principles on bone health.

8.2 Major international research projects and conferences

- An EU-COST Action (www.keelbonedamage.eu, CA15224) will continue until November 2020 and is an excellent source to provide the latest information regarding ongoing activities, workshops and data.
- Several large EU-funded projects involving assessment of hen health and welfare include ChickenStress (www.chickenstress.edu) and the 'Poultry and pig low input and organic production systems' Welfare'.

9 References

Aerni, V., Brinkhof, M. W. G., Wechsler, B., Oester, H. and Fröhlich, E. (2005). Productivity and mortality of laying hens in aviaries: a systematic review, *World's Poultry Science Journal*. Cambridge University Press on behalf of World's Poultry Science Association 61(1), 130–142.

Afazeli, H., Jafari, A., Rafiee, S. and Nosrati, M. (2014). An investigation of biogas production potential from livestock and slaughterhouse wastes, *Renewable and Sustainable Energy Reviews*. Elsevier 34, 380–386.

Ali, A. B. A., Campbell, D. L., Karcher, D. M. and Siegford, J. M. (2016). Influence of genetic strain and access to litter on spatial distribution of 4 strains of laying hens in an aviary system, *Poultry Science* 95(11), 2489–2502. doi: 10.3382/ps/pew236.

Anderson, M. and Fritsche, K. L. (2002). (n-3) Fatty acids and infectious disease resistance, *Journal of Nutrition*. 132(12), 3566–3576. Available at: http://www.ncbi.nlm.nih.gov/pubmed/12468590.

Aruoma, O. I. (1998). Free radicals, oxidative stress, and antioxidants in human health and disease, *Journal of the American Oil Chemists' Society*. Springer-Verlag 75(2), 199–212. doi: 10.1007/s11746-998-0032-9.

Bain, M. M., Nys, Y. and Dunn, I. C. (2016). Increasing persistency in lay and stabilising egg quality in longer laying cycles. What are the challenges?, *British Poultry Science*. Taylor & Francis 57(3), 330–338.

Baker, S. L., Robison, C. I., Karcher, D. M., Toscano, M. J. and Makagon, M. M. (2020). Keel impacts and associated behaviors in laying hens. *Applied Animal Behaviour Science* 222:104886 Available at: https://linkinghub.elsevier.com/retrieve/pii/S0168159119301467.

Bass, S. L. (2000). The prepubertal years: a uniquely opportune stage of growth when the skeleton is most responsive to exercise?, *Sports Medicine* 30(2), 73–78.

Baur, S. (2017). Radiographic evaluation of keel bone damage. Universität Bern.

Baur, S., Rufener, C., Toscano, M. J. and Geissbühler, U. (2020). Radiographic evaluation of keel bone damage in laying hens - morphologic and temporal observations in a longitudinal study, *Frontiers in Veterinary Science* 7, 129.

Bestman, M. and Wagenaar, J. P. (2014). Health and welfare in Dutch organic laying hens, *Animals*. Multidisciplinary Digital Publishing Institute 4(2), 374–390. doi: 10.3390/ani4020374.

Blokhuis, H., et al. (2005). Welfare aspects of various systems of keeping laying hens, *EFSA Journal* 197, 1–23.

Brantsæter, M., Nordgreen, J., Rodenburg, T. B., Tahamtani, F. M., Popova, A. and Janczak, A. M. (2016). Exposure to increased environmental complexity during rearing reduces fearfulness and increases use of three-dimensional space in laying hens (Gallus gallus domesticus), *Frontiers in Veterinary Science*. Frontiers Media SA 3, 14.

Buckner, G. D., Insko, W. M., Henry, A. H. and Wachs, E. F. (1948). Rate of growth and calcification of the sternum of male and female New Hampshire chickens, *Poultry Science*. Oxford University Press: Oxford, UK 27(4), 430–433.

Buckner, G. D., Insko, W. M., Henry, A. H. and Wachs, E. F. (1949). Rate of growth and calcification of the sternum of male and female New Hampshire chickens having crooked keels, *Poultry Science*. Oxford University Press, Oxford, UK 28(2), 289–292.

Budgell, K. L. and Silversides, F. G. (2004). Bone breakage in three strains of end-of-lay hens, *Canadian Journal of Animal Science*. Agricultural Institute of Canada 84(4), 745–747. doi: 10.4141/A04-040.

Calder, P. C. (2006). Polyunsaturated fatty acids and inflammation, *Prostaglandins, Leukotrienes and Essential Fatty Acids* 75(3), 197–202. doi: 10.1016/j.plefa.2006.05.012.

Calder, P. C. (2009). Polyunsaturated fatty acids and inflammatory processes: new twists in an old tale, *Biochimie*. 91(6), 791–795. Available at: http://www.ncbi.nlm.nih.gov/entrez/query.fcgi?cmd=Retrieve&db=PubMed&dopt=Citation&list_uids=19455748.

Campbell, D. L. M., Makagon, M. M., Swanson, J. C., and Siegford, J. M. (2015). Litter use by laying hens in a commercial aviary: dust bathing and piling. *Poultry Science* 95, 164–175.

Candelotto, L., Stratmann, A., Gebhardt-Henrich, S. G., Rufener, C., van de Braak, T. and Toscano, M. J. (2017). Susceptibility to keel bone fractures in laying hens and the role of genetic variation, *Poultry Science* 96(10), 3517–3528. doi: 10.3382/ps/pex146.

Candelotto, L., et al. (n.d.). Genetic variation of keel and long bone skeletal properties for five lines of laying hens, *Journal of Applied Poultry Research*.

Casey-Trott, T., et al. (2015). Methods for assessment of keel bone damage in poultry, *Poultry Science* 71(3), 461–472. doi: 10.3382/ps/pev223.

Casey-Trott, T. M., Guerin, M. T., Sandilands, V., Torrey, S. and Widowski, T. M. (2017a). Rearing system affects prevalence of keel-bone damage in laying hens: a longitudinal study of four consecutive flocks, *Poultry Science* 96(7), 2029–2039. doi: 10.3382/ps/pex026.

Casey-Trott, T. M., Korver, D. R., Guerin, M. T., Sandilands, V., Torrey, S. and Widowski, T. M. (2017b). Opportunities for exercise during pullet rearing, Part I: effect on the musculoskeletal characteristics of pullets, *Poultry Science* 96(8), 2509–2517. doi: 10.3382/ps/pex059.

Casey-Trott, T. M., Korver, D. R., Guerin, M. T., Sandilands, V., Torrey, S. and Widowski, T. M. (2017c). Opportunities for exercise during pullet rearing, Part II: long-term effects on bone characteristics of adult laying hens at the end-of-lay, *Poultry Science* 96(8), 2518–2527. doi: 10.3382/ps/pex060.

Casey-Trott, T. M. and Widowski, T. M. (2016). Behavioral differences of laying hens with fractured keel bones within furnished cages, *Frontiers in Veterinary Science* 3, 42.

Chargo, N. J., Robison, C. I., Akaeze, H. O., Baker, S. L., Toscano, M. J., Makagon, M. M. and Karcher, D. M. (2018a). Keel bone differences in laying hens housed in enriched colony cages, *Poultry Science*. Poultry Science Association, Inc. 98(2), 1031–1036.

Chargo, N. J., Robison, C. I., Baker, S. L., Toscano, M. J., Makagon, M. M. and Karcher, D. M. (2018b). Keel bone damage assessment: consistency in enriched colony laying hens, *Poultry Science*. Poultry Science Association, Inc. 98(2), 1017–1022.

Colson, S., Arnould, C. and Michel, V. (2008). Influence of rearing conditions of pullets on space use and performance of hens placed in aviaries at the beginning of the laying period, *Applied Animal Behaviour Science* 111(3–4), 286–300. doi: 10.1016/j.applanim.2007.06.012.

Craig, L. E., Dittmer, K. E. and Thompson, K. G. (2016). Bones and joints. In: Maxie, G. (Ed.) *Jubb, Kennedy & Palmer's Pathology of Domestic Animals*, Volume 1, 6th edn. Elsevier, St Louis, MO, pp. 16–163.

Dacke, C. G., et al. (1993). Medullary bone and avian calcium regulation, *Journal of Experimental Biology*. The Company of Biologists Ltd, 184(1), 63–88.

Darwin, C. R. (1868). Variation of plants and animals under domestication.

Del Hoyo, J., Elliott, A. and Sargatal, J. (2001). *Hand Book of the Birds of the World. Volume 2: New World Vultures to Guinea Fowl*. Lynx Editions, Barcelona.

Dunn, I. C. and Sharp, P. J. (1992). The effect of photoperiodic history on egg laying in dwarf broiler hens, *Poultry Science*. Oxford University Press, Oxford, UK 71(12), 2090–2098.

Dunn, I. C., Fleming, R. H., McCormack, H. A., Morrice, D., Burt, D. W., Preisinger, R. and Whitehead, C. C. (2007). A QTL for osteoporosis detected in an F2 population derived from White Leghorn chicken lines divergently selected for bone index, *Animal Genetics*. Wiley Online Library 38(1), 45–49.

EFSA Panel on Animal Health and Animal Welfare (2015). Scientific opinion on welfare aspects of the use of perches for laying hens. *EFSA Journal* 13, 4131.

Eggen, A. (2012). The development and application of genomic selection as a new breeding paradigm, *Animal Frontiers*. American Society of Animal Science 2(1), 10–15.

Enneking, S. A., Cheng, H. W., Jefferson-Moore, K. Y., Einstein, M. E., Rubin, D. A. and Hester, P. Y. (2012). Early access to perches in caged White Leghorn pullets, *Poultry Science*. Oxford University Press 91(9), 2114–2120.

Etches, R. J. (1987). Calcium logistics in the laying hen, *The Journal of Nutrition* 117(3), 619–628.

Eusemann, B. K., Baulain, U., Schrader, L., Thöne-Reineke, C., Patt, A. and Petow, S. (2018a). Radiographic examination of keel bone damage in living laying hens of different strains kept in two housing systems, *PLoS ONE*. Public Library of Science 13(5), e0194974.

Eusemann, B. K., Sharifi, A. R., et al. (2018b). Influence of a sustained release deslorelin acetate implant on reproductive physiology and associated traits in laying hens, *Frontiers in Physiology* 9, 1846.

FAWC (2010). *Opinion on Osteoporosis and Bone Fractures in Laying Hens*. Farm Animal Welfare Council, London. Available at: http://www.fawc.org.uk/pdf/bone-strength-opinion-101208.pdf.

FAWC (2013). *An Open Letter to Great Britain Governments: Keel Bone Fracture in Laying Hens*. Farm Animal Welfare Council, London.

Fleming, R. H., et al. (2004). Assessing bone mineral density in vivo: digitized fluoroscopy and ultrasound, *Poultry Science* 83(2), 207–214. Available at: http://www.ncbi.nlm.nih.gov/pubmed/14979571.

Fleming, R. H., McCormack, H. A., McTeir, L. and Whitehead, C. C. (2006). Relationships between genetic, environmental and nutritional factors influencing osteoporosis in laying hens, *British Poultry Science*. Taylor & Francis 47(6), 742–755.

Freeman, K. T., et al. (2008). A fracture pain model in the rat: adaptation of a closed femur fracture model to study skeletal pain, *Anesthesiology: The Journal of the American Society of Anesthesiologists*. The American Society of Anesthesiologists, 108(3), pp. 473–483.

Freire, R. and Cheng, H. W. (2004). Experience-dependent changes in the hippocampus of domestic chicks: a model for spatial memory, *European Journal of Neuroscience* 20(4), 1065–1068. doi: 10.1111/j.1460-9568.2004.03545.x.

Fröhlich, E. K. F. and Oester, H. (2001). From battery cages to aviaries: 20 years of Swiss experiences, In: Sambraus, M. G. H. and Steiger, A. (Eds.), *Das Wohlergehen von Legehennen in Europa-Berichte, Analysen und Schlussfolgerungen*. Citeseer, 28:176.

Fulton, J. E. (2012). Genomic selection for poultry breeding, *Animal Frontiers*. American Society of Animal Science, 2(1), 30–36.

Fumihito, A., Miyake, T., Sumi, S., Takada, M., Ohno, S. and Kondo, N. (1994). One subspecies of the red junglefowl (Gallus gallus gallus) suffices as the matriarchic ancestor of all domestic breeds, *Proceedings of the National Academy of Sciences of the United States of America* 91(26), 12505–12509. doi: 10.1073/pnas.91.26.12505.

Gebhardt-Henrich, S. G. and Fröhlich, E. K. F. (2015). Early onset of laying and bumblefoot favor keel bone fractures, *Animals* 5(4), 1192–1206. doi: 10.3390/ani5040406.

Gebhardt-Henrich, S. G., Pfulg, A., Fröhlich, E. K. F., Käppeli, S., Guggisberg, D., Liesegang, A. and Stoffel, M. H. (2017a). Limited associations between keel bone damage and bone properties measured with computer tomography, three-point bending test, and analysis of minerals in Swiss laying hens', *Frontiers in Veterinary Science*. Lausanne, 4(128), 128. doi: 10.3389/fvets.2017.00128.

Gebhardt-Henrich, S. G., Toscano, M. J. and Würbel, H. (2017b). Perch use by broiler breeders and its implication on health and production, *Poultry Science* 96(10), 3539–3549. doi: 10.3382/ps/pex189.

Gebhardt-Henrich, S. G., Rufener, C. and Stratmann, A. (2019). Improving intra-and inter-observer repeatability and accuracy of keel bone assessment by training with radiographs, *Poultry Science*. doi: 10.1093/ps/pez410.

Gerpe, C., Stratmann, A. and Toscano., M. J. (2017). Assessment of animal welfare during the collection process of depopulation for end of lay hens. In: de Jong, I. and Koene, P. (Eds) Proceedings of the 7th International Conference on the Assessment of Animal Welfare at Farm and Group Level. Wageningen Academic Publishers, Ede, p. 184.

Gilbert, S. F. (2000). Osteogenesis: the development of bones. In: Gilbert, S. F. (Ed.), *Developmental Biology*. 6th edn. Sinauer Associates, Sunderland, MA.

Green, P. R., Davies, I. B. and Davies, M. N. (1993). 'Interaction of visual and tactile information in the control of chicks' locomotion in the visual cliff, *Perception* 22(11), 1319-1331. doi: 10.1068/p221319.

Gregory, N. G. and Wilkins, L. J. (1989). Broken bones in domestic fowl: handling and processing damage in end-of-lay battery hens, *British Poultry Science* 30(3), 555-562. Available at: http://www.ncbi.nlm.nih.gov/entrez/query.fcgi?cmd=Retrieve&db=PubMed&dopt=Citation&list_uids=2819499.

Gregory, N. G., Wilkins, L. J., Eleperuma, S. D., Ballantyne, A. J. and Overfield, N. D. (1990). Broken bones in domestic fowls: effect of husbandry system and stunning method in end-of-lay hens, *British Poultry Science*. Taylor & Francis 31(1), 59-69.

Gregory, N. G., Wilkins, L. J., Alvey, D. M. and Tucker, S. A. (1993). Effect of catching method and lighting intensity on the prevalence of broken bones and on the ease of handling of end-of-lay hens, *The Veterinary Record* 132(6), 127-129.

Griel, A. E., Kris-Etherton, P. M., Hilpert, K. F., Zhao, G., West, S. G. and Corwin, R. L. (2007). An increase in dietary n-3 fatty acids decreases a marker of bone resorption in humans, *Nutrition Journal* 6, 2. doi: 10.1186/1475-2891-6-2.

Gunnarsson, et al. (2000). Rearing without early access to perches impairs the spatial skills of laying hens, *Applied Animal Behaviour Science*. Netherlands 67(3), 217-228. doi: 10.1016/S0168-1591(99)00125-2.

Habig, C., et al. (2017). How bone stability in laying hens is affected by phylogenetic background and performance level, *European Poultry Science*. Ulmer 81, ISSN 1612-9199, © Verlag Eugen Ulmer, Stuttgart. doi: 10.1399/eps.2017.200.

Habinski, A. M., Caston, L. J., Casey-Trott, T. M., Hunniford, M. E. and Widowski, T. M. (2017). Animal well-being and behavior: development of perching behavior in 3 strains of pullets reared in furnished cages, *Poultry Science* 96(3), 519-529. doi: 10.3382/ps/pew377.

Harlander-Matauschek, A., Rodenburg, T. B., Sandilands, V., Tobalske, B. W. and Toscano, M. J. (2015). Causes of keel bone damage and their solutions in laying hens, *World's Poultry Science Journal* 71(3), 461-472. doi: 10.1017/S0043933915002135.

Heerkens, J., et al. (2013). Do keel bone deformations affect egg-production in end-of-lay housing hens housed in aviaries? In: Tauson, R., et al. (Eds) 9th European Poultry Conference, Uppsala, Sweden, p. 127.

Heerkens, J. L. T., Delezie, E., Rodenburg, T. B., Kempen, I., Zoons, J., Ampe, B. and Tuyttens, F. A. (2015). Risk factors associated with keel bone and foot pad disorders in laying hens housed in aviary systems, *Poultry Science*. Oxford University Press 95(3), 482-488. doi: 10.3382/ps/pev339.

Heerkens, J. L. T., Delezie, E., Ampe, B., Rodenburg, T. B. and Tuyttens, F. A. (2016). Ramps and hybrid effects on keel bone and foot pad disorders in modified aviaries for laying hens, *Poultry Science*. Oxford University Press 95(11), 2479-2488. doi: 10.3382/ps/pew157.

Heikkilä, M., Wichman, A., Gunnarsson, S. and Valros, A. (2006). Development of perching behaviour in chicks reared in enriched environment, *Applied Animal Behaviour Science*. Elsevier 99(1-2), 145-156.

Hester, P. Y., Wilson, D. A., Settar, P., Arango, J. A. and O'Sullivan, N. P. (2011). Effect of lighting programs during the pullet phase on skeletal integrity of egg-laying strains of chickens, *Poultry Science*. Oxford University Press, Oxford, UK 90(8), 1645-1651.

Hester, P. Y., Enneking, S. A., Haley, B. K., Cheng, H. W., Einstein, M. E. and Rubin, D. A. (2013). The effect of perch availability during pullet rearing and egg laying on musculoskeletal health of caged White Leghorn hens, *Poultry Science*. Oxford University Press 92(8), 1972-1980.

Hocking, P. M., Bain, M., Channing, C. E., Fleming, R. and Wilson, S. (2003). Genetic variation for egg production, egg quality and bone strength in selected and traditional breeds of laying fowl, *British Poultry Science*. Taylor & Francis 44(3), 365-373.

Hudson, H. A. et al. (1993) Histomorphometric bone properties of sexually immature and mature White Leghorn hens with evaluation of fluorochrome injection on egg production traits, *Poultry science*. Oxford University Press 72(8), 1537-1547.

Hurwitz, S. (1965). Calcium turnover in different bone segments of laying fowl, *American Journal of Physiology* 208(1), 203-207.

Janczak, A. M. and Riber, A. B. (2015). Review of rearing-related factors affecting the welfare of laying hens, *Poultry Science* 94(7), 1454-1469. doi: 10.3382/ps/pev123.

Johnson, A. L. (2015). Reproduction in the female. In: Scanes, C. G. (Ed.) *Sturkie's Avian Physiology*, 6th edn. Elsevier Academic Press, London, pp. 635-665. doi: 10.1016/B978-0-12-407160-5.00028-2.

Käppeli, S., Gebhardt-Henrich, S. G., Fröhlich, E., Pfulg, A. and Stoffel, M. H. (2011). Prevalence of keel bone deformities in Swiss laying hens, *British Poultry Science*. Lidfors, L., Blokhuis, H. J. and Keeling, L. (Eds). Taylor & Francis, Uppsala, Sweden 52(5), 531-536. doi: 10.1080/00071668.2011.615059.

Karcher, D. M. and Mench, J. A. (2018). Overview of commercial poultry production systems and their main welfare challenges. In: Mench, J. A. (Ed.) *Advances in Poultry Welfare Science*, 1st edn. Woodhead Publishing, Cambridge, pp. 3-25.

Kerschnitzki, M., Zander, T., Zaslansky, P., Fratzl, P., Shahar, R. and Wagermaier, W. (2014). Rapid alterations of avian medullary bone material during the daily egg-laying cycle, *Bone*. Elsevier B.V. 69, 109-117. doi: 10.1016/j.bone.2014.08.019.

Kestin, S. C., et al. (1992). Prevalence of leg weakness in broiler chickens and its relationship with genotype, *Veterinary Record* 131(9), 190-194. Available at: http://www.ncbi.nlm.nih.gov/entrez/query.fcgi?cmd=Retrieve&db=PubMed&dopt=Citation&list_uids=1441174.

Kettlewell, P. J. and Mitchell, M. A. (1994). Catching, handling and loading of poultry for road transportation, *World's Poultry Science Journal*. Cambridge University Press 50(1), 54-56.

Knowles, T. G. and Wilkins, L. J. (1998). The problem of broken bones during the handling of laying hens - a review, *Poultry Science*. Oxford University Press 77(12), 1798-1802.

Kozak, M., Tobalske, B., Martins, C., Bowley, S., Wuerbel, H. and Harlander-Matauschek, A. (2016). Use of space by domestic chicks housed in complex aviaries, *Applied Animal Behaviour Science*. Elsevier B.V. 181, 115-121. doi: 10.1016/j.applanim.2016.05.024.

Kristensen, H. H., Berry, P. S. and Tinker, D. B. (2001). Depopulation systems for spent hens–a preliminary evaluation in the United Kingdom, *The Journal of Applied Poultry Research*. Oxford University Press 10(2), 172-177.

Lacy, M. P. and Czarick, M. (1998). Mechanical harvesting of broilers, *Poultry Science*. Oxford University Press 77(12), 1794-1797.

Lambe, N. R., Scot, G. B., and Hitchcock, D. (1997). Behaviour of laying hens negotiating perches at different heights. *Animal Welfare* 6, 29–41.

Lanyon, L. E. (1992). Control of bone architecture by functional load bearing, *Journal of Bone and Mineral Research* 7 (Suppl. 2), S369–S375. doi: 10.1002/jbmr.5650071403.

Lay, D. C., Fulton, R. M., Hester, P. Y., Karcher, D. M., Kjaer, J. B., Mench, J. A., Mullens, B. A., Newberry, R. C., Nicol, C. J., O'Sullivan, N. P. and Porter, R. E. (2011). Hen welfare in different housing systems, *Poultry Science* 90(1), 278–294. doi: 10.3382/ps.2010-00962.

LeBlanc, C. (2016). Development of domestic fowl locomotion over inclined surfaces and use of anticipation strategies. Thesis. University of Guelph. Available at: http://hdl.handle.net/10214/9670.

Leyendecker, M., Hamann, H., Hartung, J., Kamphues, J., Neumann, U., Sürie, C. and Distl, O. (2005). Keeping laying hens in furnished cages and an aviary housing system enhances their bone stability, *British Poultry Science*. Taylor & Francis 46(5), 536–544.

Liu, D., et al. (2003). Long-term supplementation of various dietary lipids alters bone mineral content, mechanical properties and histological characteristics of Japanese quail, *Poultry Science* 82(5), 831–839. Available at: http://www.ncbi.nlm.nih.gov/entrez/query.fcgi?cmd=Retrieve&db=PubMed&dopt=Citation&list_uids=12762407.

Mach, D. B., Rogers, S. D., Sabino, M. C., Luger, N. M., Schwei, M. J., Pomonis, J. D., Keyser, C. P., Clohisy, D. R., Adams, D. J., O'Leary, P. and Mantyh, P. W. (2002). Origins of skeletal pain: sensory and sympathetic innervation of the mouse femur, *Neuroscience*. Elsevier 113(1), 155–166.

Mellor, D. J., Cook, C. J. and Stafford, K. J. (2000). Quantifying some responses to pain as a stressor, *The Biology of Animal Stress: Basic Principles and Implications for Welfare*. CABI Publishing, Wallingford, Oxon, UK, pp. 171–198.

Michel, V. and Huonnic, D. (2003). A comparison of welfare, health and production performance of laying hens reared in cages or in aviaries, in *Abstracts from the Spring Meeting of the WPSA French Branch*, *British Poultry Science* 44(5), 769–831, doi: 10.1080/00071668.2003.10871381.

Mitchell, M. A. and Kettlewell, P. J. (2009). Welfare of poultry during transport – a review. In: Poultry Welfare Symposium. Association Proceeding, Cervia, pp. 90–100.

Moinard, C., Rutherford, K. M. D., Haskell, M. J., McCorquodale, C., Jones, R. B. and Green, P. R. (2005). Effects of obstructed take-off and landing perches on the flight accuracy of laying hens. *Applied Animal Behaviour Science* 93, 81–95. Available at: http://www.sciencedirect.com/science/article/pii/S0168159104003077.

Moinard, C., Statham, P., Haskell, M. J., McCorquodale, C., Jones, R. B. and Green, P. R. (2004). Accuracy of laying hens in jumping upwards and downwards between perches in different light environments, *Applied Animal Behaviour Science* 85(1–2), 77–92. doi: 10.1016/j.applanim.2003.08.008.

Nasr, M. A. F., Murrell, J., Wilkins, L. and Nicol, C. (2012a). The effect of keel fractures on egg production parameters, mobility and behaviour in individual laying hens, *Animal Welfare* 21(1), 127–135. Available at: http://www.ufaw.org.uk/documents/nasr.pdf.

Nasr, M. A., Nicol, C. J. and Murrell, J. C. (2012b). Do laying hens with keel bone fractures experience pain?, *PLoS ONE*. Public Library of Science 7(8), e42420. doi: 10.1371/journal.pone.0042420.

Nasr, M. A. F., Nicol, C. J., Wilkins, L. and Murrell, J. C. (2015). The effects of two non-steroidal anti-inflammatory drugs on the mobility of laying hens with keel bone fractures, *Veterinary Anaesthesia and Analgesia*. Wiley Online Library 42(2), 197-204.

Nencini, S. and Ivanusic, J. J. (2016). The physiology of bone pain. How much do we really know?, *Frontiers in Physiology*. Frontiers 7, 157.

Van Niekerk, T. and Reuvekamp, B. (1994). Husbandry factors and bone strength in laying hens. In: Proceedings of the 9th European Symposium on Poultry Welfare, Volume II. Glasgow, Scotland, pp. 133-136.

Norman, K. I., Adriaense, J. E. C. and Nicol, C. J. (2019). The impact of early structural enrichment on spatial cognition in layer chicks, *Behavioural Processes*. Elsevier 164(May), 167-174. doi: 10.1016/j.beproc.2019.05.008.

Norman, K. I., Weeks, C. A., Pettersson, I. C. and Nicol, C. J. (2018). The effect of experience of ramps at rear on the subsequent ability of layer pullets to negotiate a ramp transition, *Applied Animal Behaviour Science* 208, 92-99. doi: 10.1016/j.applanim.2018.08.007.

Over, R. and Moore, D. (1981). Spatial acuity of the chicken, *Brain Research* 211(2), 424-426. doi: 10.1016/0006-8993(81)90967-7.

Petrik, M. T., Guerin, M. T. and Widowski, T. M. (2015). On-farm comparison of keel fracture prevalence and other welfare indicators in conventional cage and floor-housed laying hens in Ontario, Canada, *Poultry Science* 94(4), 579-585. Available at: http://ps.oxfordjournals.org/content/94/4/579.abstract.

Pettersson, I. C., et al. (2017). The ability of laying pullets to negotiate two ramp designs as measured by bird preference and behaviour, *PeerJ*. United States 5, e4069. doi: 10.7717/peerj.4069.

Pickel, T., Scholz, B. and Schrader, L. (2010). Perch material and diameter affects particular perching behaviours in laying hens, *Applied Animal Behaviour Science* 127(1-2), 37-42. Available at: http://www.sciencedirect.com/science/article/pii/S0168159110002078.

Pickel, T., Schrader, L. and Scholz, B. (2011). Pressure load on keel bone and foot pads in perching laying hens in relation to perch design, *Poultry Science*. Oxford University Press 90(4), 715-724. doi: 10.3382/ps.2010-01025.

Pines, M. and Reshef, R. (2014). Poultry bone development and bone disorders. In: Scanes, C. G. (Ed.) *Sturkie's Avian Physiology*, 6th edn. Elsevier, London, pp. 367-364.

Podisi, B. K., Knott, S. A., Dunn, I. C., Burt, D. W. and Hocking, P. M. (2012). Bone mineral density QTL at sexual maturity and end of lay, *British Poultry Science*. Taylor & Francis 53(6), 763-769.

Pottgüter, R. (2016). Feeding laying hens to 100 weeks of age, *Lohmann Information* 501(1), 18-21. Available at: http://www.ltz.de/de-wAssets/docs/lohmann-information/Lohmann-Information_3.pdf.

Pourquié, O. (2004). The chick embryo: a leading model in somitogenesis studies, *Mechanisms of Development* 121(9), 1069-1079. doi: 10.1016/j.mod.2004.05.002.

Prescott, N. B., Wathes, C. M. and Jarvis, J. R. (2003). Light, vision and the welfare of poultry, *Animal Welfare* 12(2), 269-288.

Prunier, A., Mounier, L., Le Neindre, P., Leterrier, C., Mormède, P., Paulmier, V., Prunet, P., Terlouw, C. and Guatteo, R. (2013). Identifying and monitoring pain in farm animals: a review, *Animal*. Cambridge University Press 7(6), 998-1010.

Rath, N. C., Huff, G. R., Huff, W. E. and Balog, J. M. (2000). Factors regulating bone maturity and strength in poultry, *Poultry Science* 79(7), 1024-1032. Available at: http://ps.fass.org/content/79/7/1024.abstract.

Reed, H. J., Wilkins, L. J., Austin, S. D. and Gregory, N. G. (1993). The effect of environmental enrichment during rearing on fear reactions and depopulation trauma in adult caged hens, *Applied Animal Behaviour Science*. Elsevier 36(1), 39-46. doi: 10.1016/0168-1591(93)90097-9.

Regmi, P., Deland, T. S., Steibel, J. P., Robison, C. I., Haut, R. C., Orth, M. W. and Karcher, D. M. (2015). Effect of rearing environment on bone growth of pullets, *Poultry Science* 94(3), 502-511. doi: 10.3382/ps/peu041.

Regmi, P., Nelson, N., Steibel, J. P., Anderson, K. E. and Karcher, D. M. (2016a). Comparisons of bone properties and keel deformities between strains and housing systems in end-of-lay hens, *Poultry Science*. Poultry Science Association, Inc. 95(10), 2225-2234.

Regmi, P., Smith, N., Nelson, N., Haut, R. C., Orth, M. W. and Karcher, D. M. (2016b). Housing conditions alter properties of the tibia and humerus during the laying phase in Lohmann white Leghorn hens, *Poultry Science*. Poultry Science Association, Inc. 95(1), 198-206.

Rentsch, A. K., et al. (2019). Laying hens' mobility is impaired by keel bone fractures though effect is not reversed by paracetamol treatment, *Applied Animal Behaviour Science*. Elsevier 217, 48-56. doi: 10.1016/J.APPLANIM.2019.04.015.

Riber, A. B., Casey-Trott, T. M. and Herskin, M. S. (2018). The influence of keel bone damage on welfare of laying hens', *Frontiers in Veterinary Science*. Frontiers 5, 6.

Riber, A. B. and Hinrichsen, L. K. (2016). Keel-bone damage and foot injuries in commercial laying hens in Denmark, *Animal Welfare*. Universities Federation for Animal Welfare 25(2), 179-184.

Richards, G. J. J., Wilkins, L. J., Knowles, T. G., Booth, F., Toscano, M. J., Nicol, C. J. and Brown, S. N. (2012). Pop hole use by hens with different keel fracture status monitored throughout the laying period, *Veterinary Record* 170(19), 494. doi: 10.1136/vr.100489.

Rodenburg, T. B., et al. (2008). Welfare assessment of laying hens in furnished cages and non-cage systems : an on-farm comparison, *Animal Welfare* 17, 363-373.

Rufener, C., Baur, S., Stratmann, A. and Toscano, M. J. (2018a). Keel bone fractures affect egg laying performance but not egg quality in laying hens housed in a commercial aviary system, *Poultry Science* 98(4), 1589-1600. doi: 10.3382/ps/pey544.

Rufener, C., et al. (2018b). Subjective but reliable: assessing radiographs of keel bone fractures in laying hens with a tagged visual analogue scale, *Frontiers in Veterinary Science* 5, 124. doi: 10.3389%2Ffvets.2018.00124.

Rufener, C., Abreu, Y., Asher, L., Berezowski, J. A., Maximiano Sousa, F., Stratmann, A. and Toscano, M. J. (2019). Keel bone fractures are associated with individual mobiltiy of laying hen in aviary systems, *Applied Animal Behaviour Science* 217, 48-56. doi: 10.1016/j.applanim.2019.05.007.

Rufener, C., Rentsch, A. K., Stratmann, A. and Toscano, M. J. (2020). Perch positioning affects both laying hen locomotion and forces experienced at the keel. *Animals* 10, 1223.

Sandilands, V., Moinard, C. and Sparks, N. H. C. (2009). Providing laying hens with perches: fulfilling behavioural needs but causing injury?, *British Poultry Science*. Taylor & Francis 50(4), 395-406. doi: 10.1080/00071660903110844.

Scholz, B., Kjaer, J. B. and Schrader, L. (2014). Analysis of landing behaviour of three layer lines on different perch designs, *British Poultry Science*. Taylor and Francis 55(4), 419-426. Taylor and Francis. doi: 10.1080/00071668.2014.933175.

Schwartzkopf-Genswein, K. S., Faucitano, L., Dadgar, S., Shand, P., González, L. A. and Crowe, T. G. (2012). Road transport of cattle, swine and poultry in North America and its impact on animal welfare, carcass and meat quality: a review, *Meat Science*. Elsevier 92(3), 227–243.

Scott, G. B. (1993). Poultry handling: a review of mechanical devices and their effect on bird welfare, *World's Poultry Science Journal*. Cambridge University Press 49(1), 44–57.

Scott, G. B. and MacAngus, G. (2004). The ability of laying hens to negotiate perches of different materials with clean or dirty surfaces, *Animal Welfare* 13(3), 361-365.

Scott, G. B. and Parker, C. A. L. (1994). The ability of laying hens to negotiate between horizontal perches. *Applied Animal Behaviour Science* 42, 121-127. Available at: http://www.sciencedirect.com/science/article/pii/016815919490152X.

Scott, G. B., Lambe, N. R. and Hitchcock, D. (1997). Ability of laying hens to negotiate horizontal perches at different heights, separated by different angles. *British Poultry Science* 38(1), 48–54. https://doi.org/10.1080/00071669708417939.

Shen, C. L., Yeh, J. K., Rasty, J., Chyu, M. C., Dunn, D. M., Li, Y. and Watkins, B. A. (2007). Improvement of bone quality in gonad-intact middle-aged male rats by long-chain n-3 polyunsaturated fatty acid, *Calcified Tissue International* 80(4), 286-293. doi: 10.1007/s00223-007-9010-8.

Silversides, F. G., Korver, D. R. and Budgell, K. L. (2006). Effect of strain of layer and age at photostimulation on egg production, egg quality, and bone strength, *Poultry Science*. Oxford University Press, Oxford, UK 85(7), 1136–1144.

Simkiss, K. (1967). *Calcium in Reproductive Phsysiology*. Chapman and Hall Ltd, London.

Sisson, S., Grossman, J. D. and Getty, R. (1975). In: Getty, R. (Ed.) *Sisson and Grossman's the Anatomy of the Domestic Animals*, 5th edn. Saunders, Philadelphia, PA.

Stratmann, A., Fröhlich, E. K. F., Gebhardt-Henrich, S. G., Harlander-Matauschek, A., Würbel, H. and Toscano, M. J. (2015a). Modification of aviary design reduces incidence of falls, collisions and keel bone damage in laying hens, *Applied Animal Behaviour Science* 165, 112-123. doi: 10.1016/j.applanim.2015.01.012.

Stratmann, A., Fröhlich, E. K. F., Harlander-Matauschek, A., Schrader, L., Toscano, M. J., Würbel, H. and Gebhardt-Henrich, S. G. (2015b). Soft perches in an aviary system reduce incidence of keel bone damage in laying hens, *PLOS ONE*. Public Library of Science 10(3), e0122568. doi: 10.1371/journal.pone.0122568.

Stratmann, A., Mühlemann, S., Vögeli, S. and Ringgenberg, N. (2019). Frequency of falls in commercial aviary-housed laying hen flocks and the effects of dusk phase length, *Applied Animal Behaviour Science*. Elsevier 216, 26-32. doi: 10.1016/j.applanim.2019.04.008.

Struelens, E. and Tuyttens, F. A. M. (2009). Effects of perch design on behaviour and health of laying hens, *Animal Welfare* 18, 533-538.

Sun, D., Krishnan, A., Zaman, K., Lawrence, R., Bhattacharya, A. and Fernandes, G. (2003). Dietary n-3 fatty acids decrease osteoclastogenesis and loss of bone mass in ovariectomized mice, *Journal of Bone and Mineral Research* 18(7), 1206-1216. doi: 10.1359/jbmr.2003.18.7.1206.

Tahamtani, F. M., Nordgreen, J., Nordquist, R. E. and Janczak, A. M. (2015). Early life in a barren environment adversely affects spatial cognition in laying hens (Gallus gallus domesticus), *Frontiers in Veterinary Science*. Frontiers Media SA 2, 3.

Tallarico, R. B. and Farrell, W. M. (1964). Studies of visual depth perception: an effect of early experience on chicks on a visual cliff, *Journal of Comparative and Physiological Psychology* 17, 1695–1702. doi: 10.1046/j.1460-9568.2003.02593.x.

Tarlton, J. F. F., Wilkins, L. J., Toscano, M. J., Avery, N. C. and Knott, L. (2013). Reduced bone breakage and increased bone strength in free range laying hens fed omega-3 polyunsaturated fatty acid supplemented diets, *Bone* 52(2), 578–586. doi: 10.1016/j.bone.2012.11.003.

Tauson, R. and Abrahamsson, P. (1994). Foot and skeletal disorders in laying hens: effects of perch design, hybrid, housing system and stocking density, *Acta Agriculturae Scandinavica, Section A - Animal Science*. Taylor & Francis 44(2), 110–119. doi: 10.1080/09064709409410189.

Taylor, P. E., Scott, G. B. and Rose, P. (2003). The ability of domestic hens to jump between horizontal perches: effects of light intensity and perch colour, *Applied Animal Behaviour Science*. Elsevier 83(2), 99–108.

Thiele, H.-H. (2015). Optimal calicum supply for laying hens, *Lohmann Information* 50(2), 32–35. Available at: http://www.ltz.de/de-wAssets/docs/lohmann-information/Lohmann-Information2_2015_Vol.-49-2-October-2015_Thiele.pdf.

Thiruvenkadan, A. K., Panneerselvam, S. and Prabakaran, R. (2010). Layer breeding strategies: an overview, *World's Poultry Science Journal*. Cambridge University Press 66(3), 477–502.

Toscano, M. J., Wilkins, L. J., Millburn, G., Thorpe, K. and Tarlton, J. F. (2013). Development of an ex vivo protocol to model bone fracture in laying hens resulting from collisions, *PLoS ONE* Witten, P. E. (Ed.). Public Library of Science, 8(6), e66215. doi: 10.1371/journal.pone.0066215.

Toscano, M. J., Booth, F., Richards, G., Brown, S., Karcher, D. and Tarlton, J. (2018). Modeling collisions in laying hens as a tool to identify causative factors for keel bone fractures and means to reduce their occurrence and severity, *PLoS ONE*. Yildirim, A. (Ed.) 13(7), e0200025. doi: 10.1371/journal.pone.0200025.

Toscano, M. J., Booth, F., Wilkins, L. J., Avery, N. C., Brown, S. B., Richards, G. and Tarlton, J. F. (2015). The effects of long (C20/22) and short (C18) chain omega-3 fatty acids on keel bone fractures, bone biomechanics, behaviour and egg production in free range laying hens, *Poultry Science* 94(5), 823–835. doi: 10.3382/ps/pev048.

Toscano, M. J., Dunn, I. C., Christensen, J.-P., Petow, S., Kittelsen, K. and Ulrich, R. 2020. Explanations for keel bone fractures in laying hens: considering alternatives other than high egg production, *Poultry Science*. doi:10.1016/j.psj.2020.05.035.

Toscano, M. J. J., Wilkins, L. J. J. and Tarlton, J. F. F. (2012). Impact of a mixed chain length omega-3 fatty acid diet on production variables in a commercial free-range laying hens, *British Poultry Science* 53(3), 360–365. doi: 10.1080/00071668.2012.698726.

Vallortigara, G., Andrew, R. J., Sertori, L. and Regolin, L. (1997). Sharply timed behavioral changes during the first 5 weeks of life in the domestic chick (Gallus gallus), *Bird Behavior* 12(1), 29–40. doi: 10.3727/015613897797141290.

Vallortigara, G. and Rogers, L. J. (2005). Survival with an asymmetrical brain: advantages and disadvantages of cerebral lateralization', *Behavioral and Brain Sciences*. UC Davis, 28(4), 575–589. doi: 10.1017/S0140525X05000105.

Van De Velde, J. P., Vermeiden, J. P. W. and Bloot, A. M. (1985). Medullary bone matrix formation, mineralization, and remodeling related to the daily egg-laying cycle of Japanese quail: a histological and radiological study, *Bone* 6(5), 321–327. doi: 10.1016/8756-3282(85)90322-9.

Walk, R. D. and Gibson, E. J. (1961). A comparative and analytical study of visual depth perception, *Psychological Monographs: General and Applied* 75(15), 1-44. doi: 10.1037/h0093827.

Warren, D. C. D. E. (1937). Physiological and genetic studies of crooked keels in chickens, *Kansas Agricultural Experimental Station Technical Bulletin* 44, 1-32.

Watkins, B. A., Li, Y., Lippman, H. E. and Feng, S. (2003). Modulatory effect of omega-3 polyunsaturated fatty acids on osteoblast function and bone metabolism, *Prostaglandins, Leukotrienes and Essential Fatty Acids* 68(6), 387-398.

Webster, A. B. (2004). Welfare implications of avian osteoporosis, *Poultry Science* 83(2), 184-192. Available at: http://www.ncbi.nlm.nih.gov/entrez/query.fcgi?cmd=Retrieve&db=PubMed&dopt=Citation&list_uids=14979568.

Weeks, C. A. and Nicol, C. J. (2006). Behavioural needs, priorities and preferences of laying hens, *World's Poultry Science Journal*. Cambridge University Press 62(2), 296-307. Available at: http://journals.cambridge.org/abstract_S0043933906000195 (Accessed: 14 March 2014).

Whitehead, C. C. (2004). Overview of bone biology in the egg-laying hen, *Poultry Science* 83(2), 193-199. Available at: http://www.ncbi.nlm.nih.gov/pubmed/14979569.

Whitehead, C. C. and Fleming, R. H. (2000). Osteoporosis in cage layers, *Poultry Science*. Oxford University Press 79(7), 1033-1041.

Wichman, A., Heikkilä, M., Valros, A., Forkman, B. and Keeling, L. J. (2007). Perching behaviour in chickens and its relation to spatial ability, *Applied Animal Behaviour Science*. Elsevier 105(1-3), 165-179.

Wilkins, L. J., McKinstry, J. L., Avery, N. C., Knowles, T. G., Brown, S. N., Tarlton, J. and Nicol, C. J. (2011). Influence of housing system and design on bone strength and keel bone fractures in laying hens, *Veterinary Record* 169(16), 414. doi: 10.1136/vr.d4831.

Wolc, A., Bednarczyk, M., Lisowski, M. and Szwaczkowski, T. (2010). Genetic relationships among time of egg formation, clutch traits and traditional selection traits in laying hens, *Journal of Animal and Feed Sciences* 19(3), 452-459.

Chapter 4

Poultry health monitoring and management: bone and skin health in broilers

Gina Caplen, University of Bristol, UK

1 Introduction

Consumer demand for cheap meat has driven the broiler industry to set ever-increasing production targets, often at the expense of the animals' wellbeing. High levels of lameness and contact dermatitis, an inflammatory skin condition, within intensively reared broiler flocks have long raised welfare concern. Reduced mobility, physical restrictions in behavioural expression and pain experience, associated with leg pathologies and skin lesions, will have negative welfare repercussions. Despite some success with interventions to improve bone and skin conditions, including sustained efforts by the genetic companies to select for improvements in leg health, some of the same problems still persist decades later. One estimate indicated that 12.5 billion broilers experience leg problems annually worldwide (FAO, 2010).

With demand for chicken meat still escalating globally, it is imperative that the risk factors for lameness and contact dermatitis are fully understood so that novel and effective management solutions can be developed and better employed. Not only would the improvement of leg and skin health directly benefit broiler welfare, but there is also potential to confer many economical benefits to the producers. Lameness is associated with higher flock morbidity and mortality (Dinev, 2009) and is one of the primary causes of on-farm culling (especially in older birds). In addition, both lameness and contact dermatitis lower growth performance indices (Knowles et al., 2008; Butterworth

http://dx.doi.org/10.19103/AS.2020.0078.17

and Haslam, 2009) and are responsible for a large proportion of carcass condemnations at the slaughter plant (Kyvsgaard et al., 2013; Hashimoto et al., 2013).

2 Leg disorders and lameness

2.1 Categories of lameness and impaired walking ability

Leg disorders encompass a broad range of pathologies and can involve any combination of tendons, joints, ligaments and bones (Bradshaw et al., 2002). Lameness can arise due to developmental bone deformities, be of an infectious origin, or be degenerative (e.g. develop as a consequence of a trauma or progressive load-bearing).

2.1.1 Gait and morphology

Over the last 60 years the morphology, walking style and locomotive capacity of commercial broilers have been profoundly altered by genetic selection for increased growth rate and body mass. Selection has specifically targeted the breast muscle, which, as a proportion of the total body mass, is now double that of traditional strains (Schmidt et al., 2009). This 'pectoral hypertrophy', in addition to greater thigh muscle and leg bone mass, and relatively short legs distinguish modern broiler strains from traditional meat breeds.

Broilers display pronounced gait alterations when compared to red junglefowl (the ancestral species), and additional differences are evident between non-lame and moderately lame individuals (Caplen et al., 2012). Compensatory gait adaptations, including the wider stance and more pronounced hip and foot rotations characteristic of broilers, develop in response to increased load (body mass), morphology (unbalanced posture), pathology and injury (Corr et al., 2003a,b). In the short term such adaptations minimise the additional energy required for movement and enable the birds to retain mobility; however, ultimately, the gait adaptations themselves are likely to be fatiguing and increase the risk of modern broilers developing progressive leg pathologies.

2.1.2 Developmental lameness

A rapid growth rate and unnaturally high body mass place extreme loads on immature bones and joints. Frequently observed skeletal abnormalities include rotational (tortional) deformities such as 'twisted legs' (unilateral angulation of tibiotarsal articulation) and 'valgus-varus' deformities (Riddell and Kong, 1992). Valgus or varus angulation is mainly associated with bilateral tibial deviation and rotation (often with gastrocnemius tendon displacement); however, some

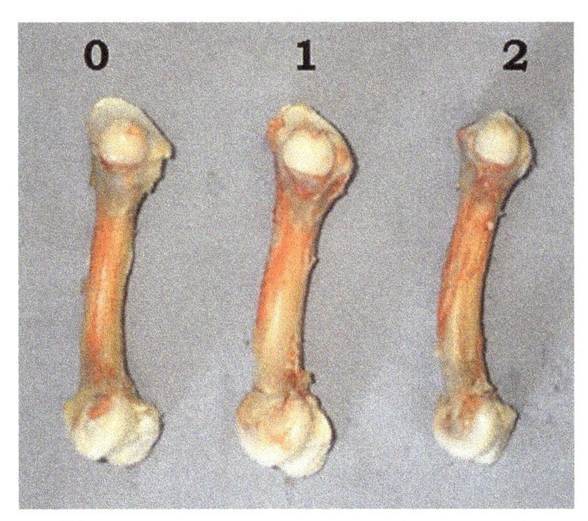

Figure 1 Examples of femoral curvature: 0 = normal, almost straight; 1 = mild curvature; 2 = moderate curvature (photo courtesy of Dr Andrew Butterworth, University of Bristol).

studies also report femoral deformations (e.g. Duff and Thorp, 1985). Such deviations remain poorly understood but are likely to be a consequence of poor bone mineralisation. Leg angulations and bone deformities can occur over a range of severities; for example valgus has been described as mild (tibia-metatarsus angle between 10° and 25°), intermediate (25–45°), and severe (>45°) (Shim et al., 2012a). A severity scale for femoral curvature is illustrated in Fig. 1.

Tibial dyschondroplasia (TD) is a leg pathology resulting from inadequate ossification and vascularisation of the epiphyseal growth plates. The TD lesion is characterised by an abnormal mass of cartilage occurring, most commonly, in the tibial metaphysis (Orth and Cook, 1994). Severe TD lesions appear to cause lameness, although the effect is less evident in milder forms (Lynch et al., 1992). Rickets is characterised by an enlargement of the epiphysial growth plate and soft rubbery bones, a result of inadequate endochondral ossification and mineralisation (Julian, 1998; Dinev and Kanakov, 2011). Rickets develops as a result of nutrient malabsorption, for example following intestinal disease, or due to inadequate nutrition, and is relatively uncommon within modern broiler flocks.

2.1.3 Infectious lameness

Bacterial chondronecrosis with osteomyelitis (BCO), more specifically known as femoral head necrosis, is recognised as a major cause of lameness in commercial broilers worldwide and is reviewed in detail elsewhere (e.g. Wideman, 2016). BCO is primarily identified by the presence of lesions in the femur (in particular, the metaphysis, femoral head and proximal femoral growth plate) and the tibiotarsus. The condition is associated with a complex range of opportunistic infective organisms, predominantly avian pathogenic *Escherichia coli* and *Staphylococcus* spp. (Dinev, 2009; Mandal et al., 2016; Al-Rubaye et al., 2017; Wijesurendra et al., 2017). Rapid increases in body weight cause mechanical damage to the immature cartilage (osteochondrosis). Bacterial proliferation at the wound sites triggers an immunological response and limits blood flow to the growth plate (Wideman and Prisby, 2013), which, in turn, leads to the development of necrotic abscesses and voids that are characteristic of BCO (Wideman, 2016). Bacteria are thought to be distributed haematologically, via the respiratory or gastrointestinal tract; *S. agnetis*, for example, can be transferred to the blood stream via drinking water (Al-Rubaye et al., 2017). Since broilers favour sitting, reduced blood flow resulting from prolonged compression of the leg arteries may further promote the development of this pathology (Wideman, 2016). Broilers housed on wire flooring are particularly susceptible to developing BCO lesions due to persistent footing instability and physiological stress (Al-Rubaye et al., 2017).

Tenosynovitis (an arthritis caused by avian reoviruses) causes the inflammation of the hock joints, lesions in the gastrocnemius and digital flexor tendons and, ultimately, lameness (Sellers, 2017). The gastrocnemius tendons leave the gastrocnemius muscle and pass over the intertarsal (hock) joint and attach to the posterior surface of the tarsometatarsus, while the digital flexor tendons extend along the tarsometatarsus to the phalanges. Upon dissection the infected synovia of swollen hocks often appears thickened, often in association with an increase in joint fluid, blood or pus (Fig. 2). In China, bacterial arthritis is also common in broilers, the main serovar being *Salmonella pullorum* (Guo et al., 2019).

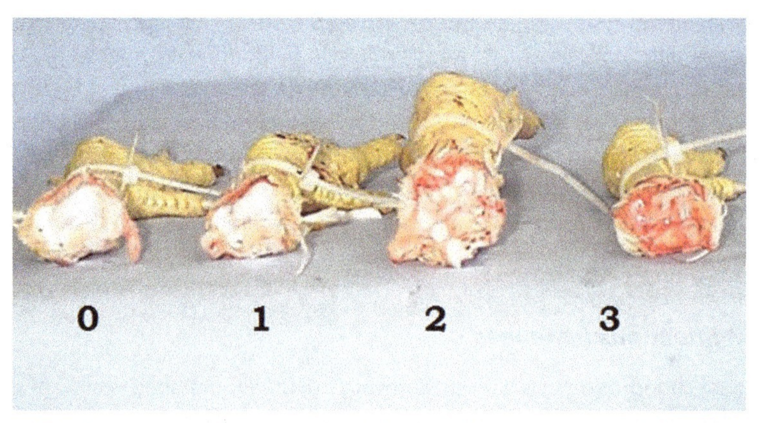

Figure 2 Inflammation of the joint capsule (proximal tarsometatarsus). Synovitis score scale: 0 = none, 1 = mild, 2 = medium, 3 = severe (photo courtesy of Dr Andrew Butterworth, University of Bristol).

2.2 Qualitative and quantitative assessment of leg health

Several gait scoring methods, based upon the visual appraisal of walking ability, have been developed for assessing broiler lameness. The widely used Bristol six-point 'gait score' (GS) scale, developed by Kestin et al. (1992), has now been adopted in a modified form for use within the broiler Welfare Quality® Assessment; scores range between 'GS 0: Normal, dextrous and agile' and 'GS 5: Incapable of walking' (Welfare Quality, 2009). Although this system is suited for on-farm welfare assessment (being relatively quick and requiring no specialised equipment), it lacks the capacity to discriminate between lameness types. A moderately lame bird assigned 'GS 3' could be affected bilaterally (e.g. valgus) or unilaterally (e.g. singular hock inflammation), or lack any obvious pathology.

Objective methodologies for quantitatively assessing gait are available, including kinetic (the measure of forces involved in walking) and kinematic (the study of body motion) systems, but at the current time these technologies remain better suited to experimental work. Several techniques have been employed to collect kinetic data from broilers, with varying measures of success. These include force plates (Corr et al., 2007; Sandilands et al., 2011), which require a constant walking speed and generate significant background 'noise', and piezoelectric pressure-sensing mats (Nääs et al., 2010). Image analysis systems have enabled the collection of three-dimensional kinematic data for the purpose of correlating gait characteristics, such as walking speed, step length and lateral body oscillation (Fig. 3), with defined lameness scores, (Caplen et al., 2012; Aydin, 2017a) and have been used to detect subtle localised changes in gait parameters as part of analgesic drug studies (Caplen et al., 2013a).

The latency-to-lie (LTL) test has been widely employed as a simple index of leg weakness (Weeks et al., 2002; Berg and Sanotra, 2003). The test is based

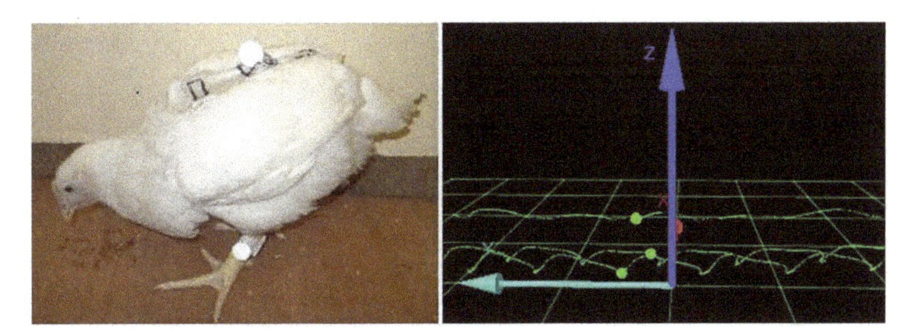

Figure 3 Broiler fitted with retro-reflective markers (one on the midline of the back, and a further two attached to the posterior aspect of the metatarsal bone, immediately above each foot). An infra-red four-camera motion capturing system was used to collect three-dimensional kinematic data as the bird walked down a runway. Images: Gina Caplen, University of Bristol.

upon the premise that chickens find sitting in water aversive and that birds with poor leg health will sit sooner when they are placed standing in shallow water (Fig. 4). Due to the logistics involved in performing this test, it is better suited to experimental trials rather than on-farm welfare assessments. An automated visual monitoring methodology has been recently developed to assign GS to broilers based upon 'the number of lying events' and 'latency to sit' when birds traverse a test corridor (Aydin et al., 2015; Aydin, 2017b). Although this new system may have the potential for transfer to a farm environment, any methodology that relies upon broilers voluntarily moving along a standard walkway will run the risk of sample bias towards the more mobile individuals (and under sample the lame birds) within a flock.

Significant efforts are currently being made to develop automated farm-based camera systems that remotely monitor flock behaviour and forewarn producers of the development of muscular-skeletal problems, for example when flock activity levels fall below an accepted level. Image analysis systems are reviewed in Chapter 7. In brief, several systems show great promise for incorporation with existing management software and wide-scale employment. Optical flow measures of flock movement have been shown to correlate with lameness (Dawkins et al., 2009, 2013; Roberts et al., 2012), while a different visual system has also been used to correlate bird activity and flock distribution with GS (Van Hertem et al., 2018).

Infrared thermography (IRT) provides a non-invasive means of measuring infrared radiation (surface heat) from an object, and the technology is being

Figure 4 The latency-to-lie test. Broilers are placed standing into shallow water at room temperature (without visual contact of conspecifics) as per the left-hand bird. The time taken for the birds to sit, as per the right-hand bird, provides a simple index of leg weakness. Birds are removed from the water as soon as they sit. Image: Gina Caplen, University of Bristol.

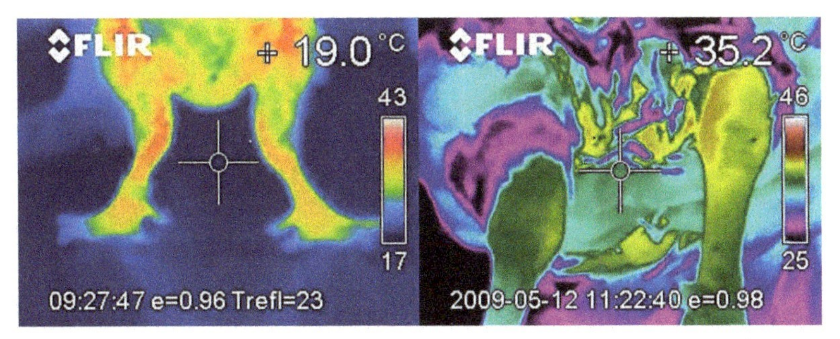

Figure 5 Thermal images of a broiler with a developmental 'valgus' leg deformity (left image) and a broiler with unilateral inflammation of the right hock joint (right image). Images: Gina Caplen, University of Bristol.

increasingly utilised in the fields of clinical and veterinary health. Although IRT cannot diagnose specific pathologies, it can be used to detect localised areas of increased or decreased heat production, that is due to inflammation or reduced blood flow, accordingly (Fig. 5). Relating to poultry production, IRT may prove to be useful as a non-invasive technique for detecting and monitoring leg pathologies on farm that are currently diagnosable only at post-mortem, for example BCO (Weimer et al., 2019).

2.3 Prevalence of lameness and specific leg pathologies

According to a comprehensive UK study, by the end of the rearing period more than a quarter (27.6%) of standard broilers had a moderate to severe gait impediment (GS 3+) and 3.3% were unable to walk (Knowles et al., 2008). More recent studies estimate a slightly lower prevalence of moderate to severe gait impediments in European flocks (Europe-wide: 15.6%, Bassler et al., 2013; Norway: 24.6%, Kittelsen et al., 2017; 19%, Granquist et al., 2019). A Danish survey found 77% of conventional broilers to have at least some form of impaired walking ability (GS 1+), while the prevalence of birds with moderate to severe lameness (GS 2+) was, a modest, 6% (Tahamtani et al., 2018).

Very little information is available regarding the current prevalence of specific leg-health pathologies. Sanotra et al. (2003) report very high levels of TD within Scandinavian flocks (Denmark: 57.1%; Sweden: 45.2–56.3%), yet later studies report much less (Finland: 2.3%, Kaukonen et al., 2017; Denmark: 5%, Tahamtani et al., 2018). BCO appears to retain a persistent presence within broiler flocks. BCO lesions have been recorded in >90% of lameness-related mortality cases (Bulgaria: Dinev, 2009), and in almost 30% of all broiler mortalities and culls (Australia: Wijesurendra et al., 2017). Although joint lesions (including arthritis and tenosynovitis) were reported to increase within UK flocks by 35% between 2011 and 2013, the actual increase was relatively small (2.16

to 2.92 per 10 000 birds slaughtered, Part et al., 2016) and, therefore, perhaps not fully reliable. Valgus and other long-bone deformities are frequently seen within standard UK broiler flocks, but exact estimates are not available. An earlier study reports a very high prevalence of valgus-varus in Scandanavia (Denmark: 36.9%; Sweden: 46.4–52.6%, Sanotra et al., 2003).

A wide variation in the estimates of general walking ability and specific leg disorders is to be expected when you consider the diversity in production systems employed on both regional and global scales (including stocking density, climate, barn design and diet). The different housing systems for broilers are described in Chapter 10. In addition, breeding companies are constantly selecting for improved production parameters and leg health (e.g. valgus/varus and TD), and so the genotype of birds supplied for use in commercial systems is in perpetual flux. In addition, management practices can become out-dated. For example, vaccination had successfully controlled tenosynovitis for decades; however, in recent years there has been a dramatic increase in the number of clinical cases as new genetic virus variants emerge (Sellers, 2017). There is an obvious need for independent and systematic monitoring of specific pathologies to document and better understand such trends.

2.4 Lameness risk factors

Key risk factors associated with lameness and poor leg health in modern broilers include those directly associated with the growth rate (including genotype, age, body mass and feeding regimen) and those indicative of sub-optimal environmental management.

2.4.1 Sex, age and body mass

Lameness increases with broiler age and body mass (e.g. Henriksen et al., 2016). Males generally have a higher GS than females and are more prone to developing valgus-varus and femoral degenerative joint lesions (Paz et al., 2013), especially if they are heavy. Some studies report a higher TD incidence in male broilers (e.g. Birgul et al., 2012), presumably since TD is also positively associated with body mass (e.g. Shim et al., 2012a). Worryingly, TD lesions have been recorded in broilers as young as 20 days, which may be a direct consequence of selection pressure for increased developmental rates (Dinev et al., 2012).

2.4.2 Genotype and system

Slower-growing genotypes demonstrate less lameness than modern fast-growing breeds when reared on the same feeding regimen (Kestin et al., 2001). Fast-growing breeds tend to have lower bone strength, which makes them susceptible to developing leg deformities (Shim et al., 2012b). The choice of system (and appropriate broiler genotype) has an obvious impact upon the leg health of the flock. Organically reared broilers tend to have better leg health than conventional flocks (Tuyttens et al., 2008; Tahamtani et al., 2018); however, flock size and stocking density are usually much lower in the organic farms, and organic systems are required to use slower-growing breeds. Fast-growing breeds do not perform well under extended production and are unsuited for use in extensive systems, as the birds make low use of the outdoor space and have a tendency to develop severe welfare problems, including high cull and mortality rates, impaired mobility, joint inflammation and severe footpad dermatitis (FPD) (Nielsen et al., 2003; Dal Bosco et al., 2014; Castellini et al., 2016).

2.4.3 Stocking density

Research consistently indicates that the health and welfare of broilers is compromised above stocking densities of 34–38 kg/m^2, dependent upon final body weight (e.g. Estevez, 2007; Sun et al., 2013; Das and Lacin, 2014). Accordingly the European Broiler Directive (2007/43/EC), which lays down the minimum rules for the protection of chickens kept for meat production, specifies that, as a general rule, stocking density should not exceed 33 kg/m^2. Although broilers grow more slowly at high stocking densities (Dawkins et al., 2004; Sun et al., 2013), there is evidence that leg health can become compromised at densities as low as 23 kg/m^2 (Buijs et al., 2009). Birds present on the barn floor effectively act as obstacles, and this has been termed the 'barrier effect' (e.g. Collins, 2008). Prolonged activity becomes more difficult at higher stocking densities, and behaviour becomes fragmented; locomotion slows and walking bouts decrease in length (e.g. Buijs et al., 2011; Ventura et al., 2012). Although stocking density does not appear to be directly linked with the prevalence of specific pathologies (e.g. valgus–varus: Arnould and Faure, 2004; TD: Das and Lacin, 2014), higher stocking densities are associated with tibial deformities and reduced bone strength (Buijs et al., 2012; Sun et al., 2013; Vargas-Galicia et al., 2017). The apparent effects of stocking density on skin health are due to correlations with deteriorating environment, especially litter quality, and this is associated with poor environmental control (Section 3.3.1). Optimal environmental management is essential for maintaining general flock health status since the combination of wet litter with warm moist air will promote bacterial growth and transmission. Gut infections, such as enteritis, provide one of many indirect mechanisms by which the intestinal microbiota may influence skeletal fitness and bone mass (Charles et al., 2015), while lameness may also arise directly as a result of bone or joint infection.

2.4.4 Environmental conditions

Key requirements for achieving and sustaining good leg health include adequate ventilation of the production facility, provision of high-quality litter and the maintenance of temperature and relative humidity (RH) within an optimum range. Higher GSs have been associated with increased hock burn (HB) and FPD scores (see Section 3), and reduced feather cleanliness, suggesting that a sub-optimal physical environment (i.e. poor litter quality) may be detrimental to leg health (Granquist et al., 2019). High temperature, high ammonia concentrations and the percentage of time that temperature and RH remain outside of the breeder's recommended range have also been associated with gait deficiencies and leg deformities (Dawkins et al., 2004; Jones et al., 2005; Tullo et al., 2017).

Rigid environmental control is also very important during incubation; temperature and oxygen concentrations can influence embryonic bone development, post-hatch production parameters and leg health (e.g. Ipek and Sozcu, 2016; Oznurlu et al., 2016). TD incidence has been associated with temperature deviations during the early stages of embryo development (Yalçin et al., 2007). Small changes in egg shell temperature (EST) during incubation could easily occur in practice due to the widespread use of large-capacity incubators and the fast growth rate of broiler embryos. Indeed, Oviedo-Rondón et al. (2009a) observed that broilers hatched from multi-stage incubators went on to develop a higher GS and greater prevalence of valgus leg deformations than those hatched from single-stage incubators.

2.5 Welfare impact of lameness

The terms 'lameness' and 'leg health' are not fully interchangeable. Although correlations between body mass, growth rate and lameness are well documented (e.g. Kestin et al., 2001), there is little evidence to conclusively link lameness severity with pathology (Garner et al., 2002; Sandilands et al., 2011; Fernandes et al., 2012). Instead, lameness is more likely to reflect the birds' subjective experience of the pathology, the manifestation of an integrated behavioural response. Investigating the true impact of poor leg health at the bird level is complicated by the gross morphological modifications that define the modern broiler bird and the multiple, often inter-linked, aetiologies and pathologies to which they are susceptible. As already mentioned lameness has been associated with contact dermatitis, and it is likely that, if left untreated, many progressive leg pathologies will confer a greater risk of developing secondary complications. Condemnations (carcasses deemed unfit for human consumption) at post-mortem inspection have been associated with increasing GS (Granquist et al., 2019), indicating that at least a proportion of lame broilers display pathological changes on the carcasses.

2.5.1 Dehydration and reduced feeding

Lame broilers may endure dehydration (Butterworth et al., 2002) since decreased mobility can limit their ability to access water, even in systems that otherwise provide appropriate water delivery. Nipple drinkers minimise water spillage and, as such, can help to maintain litter quality; however, their position, above the broilers' heads, can cause some birds to lose balance when stretching to drink (Jones et al., 2005). Very lame birds are often seen to have a lower body mass. Food intake may be reduced if it is difficult to reach; however, appetite may also become suppressed as a result of inflammatory-induced sickness behaviour (e.g. Dantzer, 2009).

2.5.2 Pain

The prevalence and severity of pain associated with broiler lameness remains poorly understood due to the impact of biomechanical factors on gait pattern, the heterogeneous nature of leg pathologies, and a requirement for rigorous experimental design (since pain experience must be inferred using indirect non-verbal measures). Strong physiological evidence indicates that broilers have the capacity to experience leg pain. Slowly adapting mechanoreceptors are present within the skin of the chicken tarsometatarsus, and these become sensitised following induced inflammation (Gentle et al., 2001). Inflammatory arthropathy, a condition that can cause pain in humans, has also been identified within the hock joints of spontaneously (naturally) lame broilers (Corr et al., 2003c). Some very lame broilers have also been reported as having a greater relative adrenal mass, which is likely to be indicative of chronic stress (Müller et al., 2015).

Visual identification of a pain state is complicated by the fact that non-lame broilers spend the majority of their time sitting (Weeks et al., 2000). In addition, poultry are a prey species and it is likely that they avoid overt pain-associated behaviour, as any display of weakness could make them more vulnerable to predation. Morphology (via mechanical limitations and a reduced motivation to walk) has a significant influence upon gait and activity levels in modern strains, regardless of any assumed discomfort. Since certain pathologies (e.g. a mild skeletal deformity) are likely to be less painful than others (e.g. inflammatory or necrotic conditions), the welfare implications of failing to quantify and differentiate pain from other causes of gait abnormality are substantial.

The provision of analgesic drugs under experimental conditions can provide indirect evidence for pathological pain if positive changes in behaviour, or improvements in a predefined test performance, are observed post-treatment. The majority of studies to date have utilised non-steroidal anti-inflammatory drugs (NSAIDs), such as carprofen and meloxicam, as they are routinely used to manage pain associated with osteoarthritis in dogs and cats and may have a therapeutic potential in poultry. A self-selection experiment, reporting the preferential selection of food spiked with NSAID, claimed to provide early evidence for the occurrence of pain in lame broilers (Danbury et al., 2000); however, a later study was unable to corroborate these findings (Siegel et al., 2011).

NSAID treatment has been observed to increase walking velocity and modify gait in moderately lame broilers (Nääs et al., 2009; Caplen et al., 2013a) and improve LTL performance (Hothersall et al., 2016). However, such improvements in mobility could be attributable to a reduction in joint inflammation as opposed to any direct analgesic effect. Although the relationship between lameness and pain (and thus welfare) is complicated by confounding 'risk factors' such as sex, strain, bodyweight and pathology, 'lameness' has been identified as the most consistent predictor for several broiler mobility measures. This indicates that there is a constituent of 'lameness' that cannot be explained by any combination of the more obvious bird characteristics (e.g. being a heavy male), and it is this component that may represent pain or discomfort (Caplen et al., 2014).

Hyperalgesia (heightened sensitivity to pain) is prevalent in many disease states as part of an inflammatory response to prevent further tissue damage. Pain-producing chemicals (cytokines and chemokines) trigger primary afferent nociceptor sensitisation. Primary hyperalgesia describes pain sensitivity that occurs directly in the damaged tissues; a lowered nociceptive/pain threshold to both thermal and mechanical stimuli is usual. Caplen et al. (2013b) utilised a specially designed apparatus to detect primary thermal hyperalgesia in broilers with experimentally induced inflammatory arthropathies (an acute pain model) and demonstrated that NSAID treatment could reverse this effect via anti-nociception (Fig. 6). A comparable effect has not yet been demonstrated in broilers with non-induced 'spontaneous' lameness; conversely, a higher baseline thermal threshold was reported in farm-lame broilers, which increased further following NSAID treatment, potentially due to altered nociceptive processing (Hothersall et al., 2014). The fact that hyperalgesia was detected in a group of experimental birds following pain induction under controlled conditions but not in a group of chickens with (potentially mixed) on-farm pathologies does not prove that the latter group had no pain experience. Pain is not always accompanied by hyperalgesia, and, therefore, a lack of response is not an evidence for an absence of pain. Clear difficulties exist in obtaining large groups of birds with comparable pathologies for experimental work and in linking lameness severity with pathology (e.g. Sandilands et al., 2011).

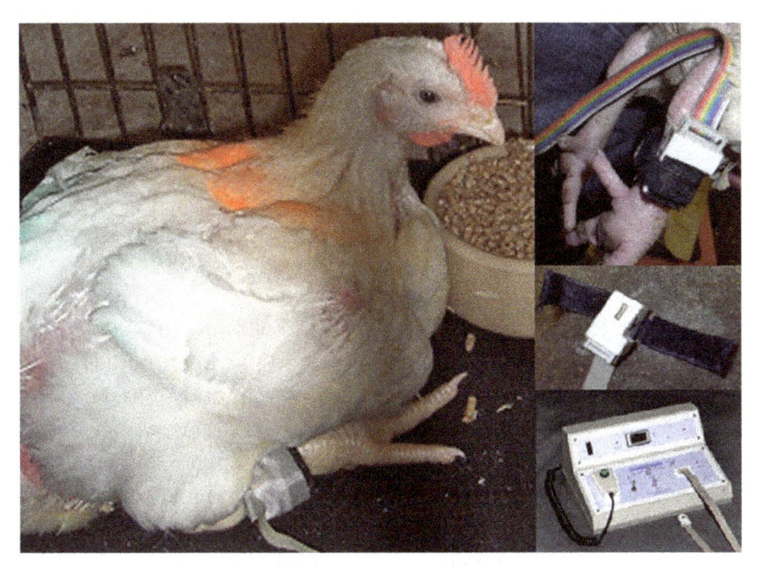

Figure 6 Apparatus used for thermal threshold testing (Images: Gina Caplen). The leg probe (containing a temperature sensor and a heated element) was attached to the lateral aspect of the tarsometatarsus via a Velcro strap. A ramped heat stimulus was applied until a behavioural endpoint was detected and then immediately removed. The response temperature was held on a digital readout.

2.5.3 Limited behavioural expression

Selection for broiler growth has significantly narrowed their ethogram and altered their time budget, compared to other breeds of *Gallus gallus domesticus,* by restricting the range of behaviours that they are physically able to perform. The effect of posture on resting metabolic rate becomes increasingly significant as broilers grow; locomotion becomes very energetically expensive, standing more so than sitting. Since the metabolic scope for exercise decreases throughout their development, the proportion of the overall metabolic rate accounted for by locomotor behaviour also decreases, which corresponds to declining activity levels and low walking speeds (Tickle et al., 2018). This is particularly apparent in the fast-growing breeds, since slower-growing broilers perform less sitting and more perching, walking and ground scratching throughout the production period (Bokkers and Koene, 2003; Reiter and Bessei, 2009).

In welfare terms inherent inactivity is problematic for several reasons. Walking acts to strengthen muscles and bones, and inactivity is thought to be a direct cause of leg weakness. Lameness further compromises mobility, limiting behavioural expression and reducing the value of any enrichment provision. There is evidence that physically impaired individuals retain at least some motivation to perform locomotory behaviour (Rutten et al., 2002; Bokkers and Koene, 2004; Bokkers et al., 2007). When this motivation remains unfulfilled, it is likely to trigger frustration (e.g. as indicated by displacement preening, Bokkers and Koene, 2003), stress and suffering.

2.6 Prevention and control of lameness

2.6.1 Genetics

Long-term selection for improved broiler leg health (in particular, long-bone deformities and TD) over the last 30 years or so has been partially effective, despite some unfavourable genetic correlations with body mass (Kapell et al., 2012a). A marked reduction in TD within commercial strains over recent years is testimony to what is possible. Estimated heritabilities of non-infectious skeletal disorders (including TD and valgus-varus), and susceptibility to infection (e.g. FHN), indicate that genetic selection and breeding programmes offer a means to further reduce multiple lameness aetiologies and leg pathologies alongside the main selection focus, to further improve production parameters (Akbaş et al., 2009; Wideman et al., 2014). At the producer level, a greater commercial uptake of slower-growing strains (with their lower predisposition for musculo-skeletal health problems) housed within appropriate systems would directly alleviate the main welfare concerns.

2.6.2 Incubation conditions

Incubation lighting and heating schedules both appear to have a marked effect upon leg health. Egg exposure to continuous light has been shown to have a detrimental effect upon embryonic leg bone development. The risk of both poor bone strength and developing TD in later life is increased, compared to incubation regimes that include a continuous period (at least 8 h) of darkness (van der Pol et al., 2017, 2019).

The optimum incubation EST for healthy leg development appears to be lower than 37.8°C, the temperature currently recommended for maximising hatchability and chick growth. Embryos incubated under 'slow start' conditions, that is under lower temperatures (EST: 36.9–37.5°C) during the first few weeks of incubation, hatched later, had greater femoral bone ash (Groves and Muir, 2014; Muir and Groves, 2018), grew slower over the first week post-hatch, and had a lower prevalence of TD lesions at day 34 (Groves and Muir, 2017). Hatchery and chick quality issues clearly influence the susceptibility of broilers to BCO. Although it remains widespread within European flocks, the pathology is being successfully addressed via improvements in hatchery hygiene (Dinev, 2009).

2.6.3 Barn management

As mentioned previously, it is very important to effectively utilise heating and ventilation systems to stringently maintain the environmental parameters of the barn within the guidelines for the breed. Maintaining dry litter also appears to be important; the use of wheat straw as litter has been associated with a higher prevalence of lameness than (more absorbent) wood-shavings (Su et al., 2000), while the addition of vermiculite to wood-shavings has been found to lower lameness severity further (Yildiz et al., 2014).

2.6.4 Diet and feeding regimes

Diet and feeding regimes are utilised to control growth rate and/or increase bone strength. Quantitative or qualitative dietary restriction can be used as a means to slow weight gain and promote healthier anatomical leg development (Kestin et al., 2001; Wijtten et al., 2010). However, feed restriction in any form will trigger overt hunger, especially in the meat breeds that have been selected for high appetite and metabolic growth. Hunger is often associated with frustration, stereotypies and adverse behaviour, occasionally including cannibalism (Eriksson et al., 2010), all of which have obvious adverse impacts upon bird welfare.

A change from ad-libitum feeding to meal feeding (i.e. limiting food availability to 240 min/day split between two, three and four discrete meal times) has been seen to improve walking ability and reduce TD (Su et al., 1999). Responsive nutritional management can also be used to control lameness by slowing down the rate at which young broilers grow and then speeding the growth rate back up to achieve a standard finishing commercial body mass once the leg bones have better developed. This can be achieved via the sequential feeding of two diets, a high-energy/low-protein diet and a low-energy/high-protein diet, over 48-hour feeding cycles (Leterrier et al., 2008); a standard finishing diet is provided from day 29 until the last day of production to compensate for reduced growth. The inclusion of whole wheat within the diet is beneficial as it slows digestion and lowers the feed conversion rate, thereby, reducing both growth rate and lameness (Knowles et al., 2008). All three of these management practices, via health improvements and decreased feed costs, have been identified as having economic potential to realise substantial improvements in gross margin and net return for the farmer (Gocsik et al., 2017).

Broilers fed mashed diet have been shown to have higher bone ash and lower GS than broilers fed the same diet in pellet form (Brickett et al., 2007). Presumably birds eat a greater volume of pelleted food (and achieve a higher corresponding weight gain), while those fed mashed diet benefit physiologically from being able to preferentially select components of their ration and benefit behaviourally from spending more time foraging for their feed. Since deficiencies and imbalances of numerous vitamins and minerals can impact upon broiler bone mineralisation and leg health, the provision of an optimal dietary formulation is extremely important. Due to the ongoing selection for improved production parameters, dietary guidelines require a regular review. Numerous nutritional factors influence broiler leg health, often via complex interactions (e.g. facilitating mineral assimilation), and these are reviewed in detail elsewhere (see Waldenstedt, 2006; Kwiatkowska et al., 2017). Calcium (Ca), phosphorus (P) and vitamin D_3 (cholecalciferol) appear to be of key importance.

Dietary Ca supplementation is beneficial for increasing bone quality and reducing TD incidence (Coto et al., 2008; Abdulla et al., 2017); sources such as oyster shell, snail shell and limestone are reported as being most effective (Oso et al., 2011). In addition to quantity, the dietary balance between available Ca and P is extremely important. The optimum feed ratio is 2:1, and an excessive supply of either element can lessen the assimilation of both (Coto et al., 2008; Bradbury et al., 2014). Rickets can occur as a Ca-deficiency or P-deficiency type, resulting from either a direct dietary deficiency or an excessive proportion of either. An increase in TD incidence has also been reported as a consequence of Ca:P imbalance (Waldenstedt, 2006). Vitamin D_3 is beneficial as a dietary supplement since it increases Ca and P intestinal absorption, improves bone quality and walking ability (Baracho et al., 2012; Sun et al., 2013) and is valuable in the prevention of TD (Whitehead et al., 2004) and BCO (Wideman et al., 2015). Ascorbic acid (vitamin C) supplementation has been found to benefit both bone quality (Yildiz et al., 2009) and walking ability (Petek et al., 2005).

Phytase is an enzyme that acts to increase the bioavailability of various minerals (e.g. Ca, P, Mg, Zn, Fe), a large proportion of which are present within grains and seeds as insoluble complexes. Dietary phytase has been shown to increase the apparent digestibility of Ca by 4–6% (Saima et al., 2009) and to improve bone mineralisation (Bradbury et al., 2017).

Dietary supplementation of probiotics (live organisms intended to improve the gut microflora), prebiotics (non-digestible feed components that promote the growth of beneficial intestinal microorganisms) and synbiotics (the combination of a probiotic with a prebiotic) benefit bone development and mineralisation by increasing the intestinal absorption and assimilation of nutrients and minerals (Scholz-Ahrens et al., 2007; Yan et al., 2018). The inclusion of a dietary synbiotic (containing a prebiotic and a probiotic mixture of four microbial strains) was found to improve multiple indices of broiler leg health, LTL performance and walking ability (Yan et al., 2019). Probiotics are most effective at reducing BCO incidence if they are given proactively as part of the feed, rather than therapeutically after the onset of lameness (Wideman et al., 2012).

It is also important to recognise that there are environmental contaminants that can negatively impact upon health, even with the provision of an 'optimum diet'. Bacterial, viral and parasitic infections can reduce the ability of the intestinal epithelium to absorb nutrients, while feed contamination with certain mycotoxins can induce or exacerbate skeletal problems due to interference with vitamin D metabolism (Waldenstedt, 2006).

2.6.5 Photoperiod and light intensity

Continuous bright light is normally provided during the first 4 days following barn placement post-hatch (the brooding period) to stimulate feeding; however, exposure to intermittent lighting during the same period has been shown to benefit leg health by slowing bone development and increasing leg bone symmetry (van der Pol et al., 2015). Within the EU the Broiler Directive (2007/43/EC) states that lighting must follow a 24-h rhythm and include periods of darkness lasting at least 6 h in total, with at least one uninterrupted period of darkness of at least 4 h, excluding dimming periods (see Chapter 10). The presence of a dark phase (scotophase) throughout the production cycle is conclusively beneficial for broiler leg health. Provision of a scotophase is associated with a reduction in lameness (Brickett et al., 2007; Knowles et al., 2008; Bassler et al., 2013; Schwean-Lardner et al., 2013; Das and Lacin, 2014) and TD (Petek et al., 2005; Karaarslan and Nazligul, 2018) and an increase in tibial strength (Lewis et al., 2009; Yang et al., 2015), compared to exposure to continuous light. Bone mineralisation peaks during the dark period and is sensitive to diurnal rhythm (Russell et al., 1984, as cited by Bassler et al., 2013). In addition, broilers provided with a scotophase are physically more active during the light period than those kept under near-continuous light (Sanotra et al., 2002; Bayram and Özkan, 2010; Schwean-Lardner et al., 2012). A photoperiod of 16L:8D appears optimal for maximising both welfare and feed conversion (Classen, 2004).

Broilers are typically housed under low light intensity during the light phase for the majority of the production cycle. Those reared under very dim lighting (<5 lux) develop a higher body mass, display less activity and have a higher GS than birds reared under brighter (10–320 lux) lighting regimes (Blatchford et al., 2009, 2012; Senaratna et al., 2016; Fidan et al., 2017). Exposure of chicks to low levels of UV-B illumination (to simulate natural daylight) has been trialled as an alternative non-dietary means of enhancing vitamin D_3. UV-B treatment has been found to have substantial benefits for bone characteristics and virtually eliminates growth-plate abnormalities caused by Ca and vitamin D deficiencies (Fleming, 2008). Broilers reared under environments including supplementary UV-A light-emitting diode (LED) and UV-B fluorescent lighting had a lower GS than those of a white LED control group (James et al., 2018). The incorporation of UV wavelengths within commercial lighting regimes may, therefore, be beneficial for broiler leg health.

2.6.6 Environmental enrichment

Increasing activity levels have known benefits for broiler leg health since mechanical loading is essential for maintaining normal bone formation and remodelling (Hester et al., 2013). Broilers encouraged to walk on treadmills were shown to have better bone density, thickness and fewer leg bone deformities than control birds (Reiter and Bessei, 2009), while cage-reared broilers were observed to have lower leg bone mineralisation than birds reared in open barn systems (Aguado et al., 2015).

Standard broiler breeds are typically barn-reared in low complexity environments with minimal enrichment; however, studies that have provided commercial flocks with elevated structures in an attempt to encourage locomotor activity (and thus improve leg and foot pad health) have met with varying success. Although the provision of platform perches in combination with dust baths was found unsuccessful in reducing lameness (Bailie et al., 2018a), the provision of platforms in a different study improved GS and TD incidence/severity (Kaukonen et al., 2017). Presumably the key to use is in the platform design. A small-scale study using experimental pens reported that slow-growing breeds use perches more frequently than fast-growing strains (Bokkers and Koene, 2003), presumably due to intrinsically higher activity levels and superior leg strength; however, low usage overall suggests perch provision to be inappropriate for broilers. In contrast, the provision of straw and/or wood-shaving bales has reliably been associated with increased physical activity (Bailie and O'Connell, 2015; Ohara et al., 2015; de Jong and Gunnink, 2019) and improved leg health (Bailie et al., 2013; Vasdal et al., 2019) in commercial flocks.

Body mass appears to physically limit the utilisation of elevated structures. Usage undergoes a marked decline towards the end of the production cycle (perches: Bailie and O'Connell, 2015; hay bales: Ohara et al., 2015; platforms: Norring et al., 2016), and females exploit the enrichment more than the (heavier) males (Estevez et al., 2002; Ohara et al., 2015). The design of commercially appropriate feeding regimes (e.g. distribution of the dietary ration on the litter surface and wheat and mealworm provision in addition to the dietary ration) to stimulate activity has largely been unsuccessful (Bizeray et al., 2002a,b; Jordan et al., 2011). The provision of oat hulls has, however, been shown to promote dust-bathing and foraging and reduce lameness (Baxter et al., 2018). The provision of slow-growing breeds with an outdoor range has many fundamental welfare benefits over indoor systems, including the encouragement of increased locomotor activity (Nielsen et al., 2003;

Sosnówka-Czajka et al., 2007) and the potential for lowering lameness levels (Zhao et al., 2014; Ipek and Sozcu, 2017).

Selection for fast early growth rate has made broilers susceptible to developing leg disorders as a consequence of significant morphological alterations and a predisposition to inactivity. This, in turn, produces direct welfare problems and confers difficulties in improving housing design. Although the majority of broiler welfare problems have a genetic basis, many, such as lameness, are further exasperated by interactions with a poor environmental management. The provision and maintenance of high-quality litter is very important for general flock health. Contact dermatitis is a skin condition frequently seen in commercial broiler flocks, being particularly prevalent in lame birds since it affects those skin surfaces that have prolonged contact with poor quality (wet) litter.

3 Contact dermatitis

Broilers are particularly susceptible to developing contact dermatitis, due to management practices (particularly litter contamination with faecal ammonia and water) and low activity levels. Contact dermatitis can range in severity from dark skin discolouration and superficial erosions (mild) to pronounced inflammation of subcutaneous tissue and deep necrotic lesions (severe). Most commonly the plantar surface of the feet, including the skin of the central pad and toes, is affected, and this is referred to as footpad dermatitis (FPD) or pododermatitis. Contact dermatitis is also commonly observed to affect the skin of the hock joint and that overlying the sternum, termed 'hock burn' (HB) and breast burn, accordingly.

3.1 Prevalence

A wide variation in the occurrence of moderate to severe FPD has been reported, both between and within studies (e.g. Italy: 0–90%, Meluzzi et al., 2008a,b; USA: 5%, Opengart et al., 2018; Denmark: 13.1%, Lund et al., 2017; Bulgaria: 16%, Dinev et al., 2019; Canada: 28.7%, Hunter et al., 2017; Europe: 37.3%, Bassler et al., 2013; the Netherlands: 38%, de Jong et al., 2012a; UK: 0–48%, Jones et al., 2005; France: 83%, Allain et al., 2009). Although the majority of these studies used a three-point scale, some, for example Allain et al. (2009), used complex scoring systems which can make inter-study comparisons difficult.

Higher prevalence and severity of FPD in European, compared to American, flocks could reflect multiple factors. These include the common European practice of flock thinning (see Section 3.3.1), the type of bedding material (chopped straw litter is frequently used in some parts of Europe while wood-shavings were used in the American study), a more pronounced seasonal (winter) effect in certain parts of Europe, concrete floors in broiler houses (more likely to produce condensation and trap litter moisture compared with the dirt floors typically found in U.S. broiler houses) and the utilisation of a relatively thin layer of bedding, completely replaced between flocks (compared to the deeper build-up and re-use of litter common in the USA) (Opengart et al., 2018).

Although HB develops more slowly than FPD (Skrbic et al., 2015) and is observed less frequently, the two are often associated (e.g. Bassler et al., 2013). As with FPD, a wide variation in HB prevalence has also been reported (e.g. Bulgaria: <1%, Dinev et al., 2019; UK: 0–33%, Haslam et al., 2007; Hepworth et al., 2011; Italy: 3–87%, Melluzzi et al., 2008a,b; Europe: 7.9%, Bassler et al., 2013; France: 59%, Allain et al., 2009), presumably for similar reasons. Although the presence of breast burn and severe HB has been positively correlated (Allain et al., 2009), breast burn is much less frequently seen (e.g. UK: <1%, Haslam et al., 2007; Bulgaria: 5%, Dinev et al., 2019; France: 16%, Allain et al., 2009; Portugal: 18%, Gouveia et al., 2009) and, as a consequence, the condition remains comparatively understudied.

3.2 Qualitative assessment

Many different scoring systems have historically been used to study FPD and HB, ranging from simple binary scores to complex severity scales with as many as ten different categories (e.g. Allain et al., 2009). To enable a meaningful assessment of dermatitis prevalence over time and between systems, it is important to establish a standardised measure. Widely used scoring systems for categorising FPD and HB lesion severity in broilers on farm as part of the 'Welfare Quality (2009) Assessment Protocol for Poultry' utilise five-point severity scales and photographic reference images (Fig. 7). To account for the likelihood that FPD lesion severity is unevenly distributed over commercial broiler units, it is recommended that at least five different sampling locations are employed within each unit and a minimum of 100 birds are scored (de Jong et al., 2012b).

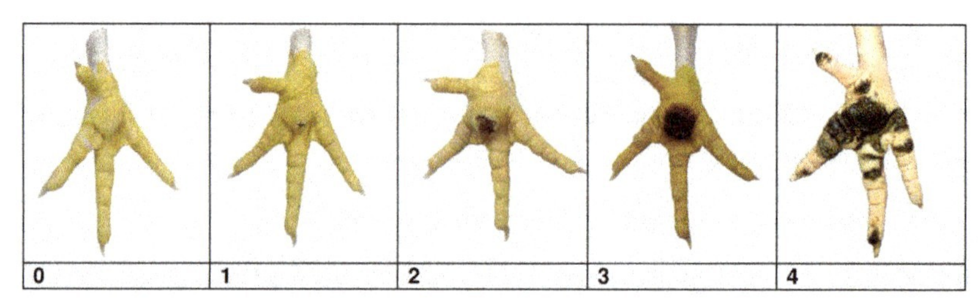

Figure 7 A five-point scoring scale (0–4) for categorising the severity of foot pad dermatitis lesion in broilers (Welfare Quality®, 2009).

3.3 Contact dermatitis risk factors

3.3.1 Age and stocking density

Older broilers are heavier, less active (e.g. Alvino et al., 2009), and consequently experience more skin contact with increasingly faecal-loaded litter. Although contact dermatitis is often reported as increasing with age (FPD: Kyvsgaard et al., 2013; Sarica et al., 2014; HB: Kjaer et al., 2006; Haslam et al., 2007), there are exceptions. FPD lesions have been observed as early as 7 days post-hatch when litter moisture is low (Berk, 2009; Hashimoto et al., 2011). This suggests that young broilers may be more vulnerable to acquiring FPD, with or without exposure to poor litter. There is further evidence for an age-related increase in skin resistance to environmental irritation. Early exposure to high-moisture litter has been shown to increase FPD prevalence and severity (at 14 days); however, no further increase was observed following litter wetting at >56 days of age (Cengiz et al., 2011).

A positive relationship between contact dermatitis and stocking density has often been reported (FPD: Haslam et al., 2007; Bailie et al., 2018b; Ventura et al., 2010; Kyvsgaard et al., 2013; HB: Karaarslan and Nazligul, 2018), associated with high litter moisture and pH (Dozier et al., 2006; Petek et al., 2010). However, stocking density is not a definitive indicator of poor welfare as several studies report no direct impact of density on FPD (Dawkins et al., 2004; Haslam et al., 2006; Meluzzi et al., 2008a; Allain et al., 2009).

Some systems stock at high density and incorporate flock thinning into their production cycle. This involves the removal of a proportion of the flock, usually 1 week prior to the end of production, in order to maintain stocking density (body mass per area) within legal limits. More severe FPD has been reported in birds at final depopulation than in younger thinned-out birds, consistent with the theory that the risk increases with age; however, litter improvement following a reduction in stocking density (post-thinning) may actually enable foot pad health to improve with age (de Jong et al., 2012a).

It has been suggested that placement density (m^2/bird) may be a more important factor than finishing density (kg/m^2), assuming other environmental factors that impact FPD remain constant. FPD lesions have been observed to form relatively early in life (between 12 and 21 days), correlating with deteriorating litter conditions (including increases in ammonia and moisture), and then to persist over the life of the flock (Opengart et al., 2018).

3.3.2 Sex and body mass

There is plenty of evidence to indicate that male and heavier birds are more at risk of developing HB (Oviedo-Rondón et al., 2009b; Henriksen et al.,

2016; Louton et al., 2018). Indeed body mass at only 2 weeks of age has been identified as a predictor of those flocks most at risk of developing HB prior to slaughter (Hepworth et al., 2010). Although some consider male birds to be more pre-disposed to developing FPD than females due to a heavier body mass (e.g. Gouveia et al., 2009; Sarica et al., 2014; Villarroel et al., 2018), not all studies are in agreement with this. Conflicting observations may reflect a loss of condition (and thus body mass) in broilers with severe FPD, especially if the lesion becomes the site of a secondary infection (e.g. Kyvsgaard et al., 2013).

3.3.3 Production systems

Contact dermatitis lesions are more prevalent/severe in the fast-growing than slower-growing genotypes (FPD: Sarica et al., 2014; Yamak et al., 2016; HB: Haslam et al., 2007; Skrbic et al., 2015), presumably due to different feed conversion rates (and related excreta properties) and activity levels. Unsurprisingly, both FPD prevalence and severity are higher in solid-floor than wire-floor (cage) housing systems (Cengiz et al., 2013; Simsek et al., 2014).

A higher prevalence of FPD has been reported in extensive indoor systems compared to traditional free-range farms (Portugal: Gouveia et al., 2009), and lower HB has been observed in organically reared broilers than in conventional flocks (Broom and Reefmann, 2005; Tuyttens et al., 2008). However, other studies have reported higher FPD scores in flocks with outdoor access than those without (UK: Pagazaurtundua and Warriss, 2006; Turkey: Sarica et al., 2014). The feet of free-range birds will be exposed to ammonia-rich litter in the barns at night, in addition to hard stony ground and pathogens (from wild birds/rodents) while they are out on the range. Broilers with access to a range incorporating both trees and grass have been reported to have better foot condition than those provided with grass alone (Dal Bosco et al., 2014), and this improvement is likely to reflect an increase in range use. Since range conditions, particularly wetness, are an additional risk factor for the pathology (Sans et al., 2014), a well-considered range design, (including the provision of shelter and drainage) should benefit foot health by allowing broilers feet to dry during their time spent outside.

3.3.4 Litter quality

The Welfare Quality Assessment Protocol for broilers describes a five-point scale for scoring litter quality ranging from '0: Completely dry and flaky, moves easily with foot' to '4: Sticks to boots once the cap or compacted crust is broken' (Welfare Quality, 2009). The prevalence and severity of contact dermatitis are often correlated with deterioration in litter quality, particularly when the surface of the litter becomes compacted and the moisture content increases (FPD: Kyvsgaard et al., 2013; de Jong et al., 2014; HB: Allain et al., 2009; Bassler et al., 2013). Wet litter and litter containing a high nitrogen content (from animal waste) are generally more alkaline, that is have a higher pH value (Meluzzi et al., 2008b; Abd El-Wahab et al., 2013). Indeed, improvements in litter quality have been shown to reduce FPD lesion severity in market-age broilers (Cengiz et al., 2011).

Litter quality is related to the type of bedding material used, nutrition, gut health and the optimal management of the housing environment (including heating, ventilation and minimising water spillage). The use of misting systems has been linked with higher levels of FPD (Jones et al., 2005), while the provision of more drinkers per unit area leads to higher litter moisture and a greater prevalence of HB (Jones et al., 2005). Increased water consumption has also been associated with higher FPD (Manning et al., 2007) and HB (Hepworth et al., 2010). Litter quality is presumably reduced directly via increased water spillage and indirectly via increased (and loose) excreta.

Several studies have highlighted that, at least under commercial conditions, litter condition is likely to be more critical than stocking density in the development of FPD (Dawkins et al., 2004; Haslam et al., 2007). Stocking density can contribute towards litter moisture levels; however, with appropriate gut health and effective environmental management, the maintenance of litter quality appears to be possible throughout the production cycle, even under high stocking densities.

3.3.5 Diet

Nutrition, including diet composition and feeding programs, plays a significant role in the aetiology of contact dermatitis due to feed conversion rate, water intake, properties of excreta and the resulting litter quality (see Section 3.3.4). Links between physiological stress, elevated corticosterone and increased water content of faeces are well established (e.g. Puvadolpirod and Thaxton, 2000; Nicol et al., 2009), most likely due to sub-optimal gut function. Jacob et al. (2016a) observed an increase in FPD prevalence between 19 and 29 days old, which correlated with a dietary switch from starter to grower feed; this abrupt dietary transition was hypothesised to reduce gut health and, therefore, litter quality. Presumably the risk of causing stress to the digestive system could easily be reduced by introducing a novel diet gradually, initially presented as a low proportion of the ration and increased over time.

Studies outlining the optimal dietary levels of crude protein, biotin and electrolytes, and the effects of diet composition and nutrient concentration on litter quality and FPD occurrence in broilers, are reviewed in detail elsewhere (e.g. Swiatkiewicz et al., 2017). Dietary constituents vary between countries, and many studies have trialled novel protein sources. FPD is significantly increased in broilers receiving an entirely vegetable-based protein diet compared to those given a diet containing both vegetable- and animal-based proteins

(Nagaraj et al., 2007; Cengiz et al., 2013), and further FPD reductions can be achieved with the addition of corn gluten meal (Eichner et al., 2007). Barley-based diets have been associated with poor litter quality and FPD increase compared to corn-based diets (Cengiz et al., 2017). On the basis of FPD prevalence, rapeseed meal has also been identified as an unsuitable alternative dietary protein to soyabean meal (Abd El-Wahab et al., 2018).

Dietary additives have also been well-researched. The provision of an increased percentage of dietary wheat has been linked with reduced HB (Haslam et al., 2007) and is also seen to benefit leg health (see Section 2.7.4). The inclusion of reduced levels of 2-hydroxy-4-(methylthio)butanoic acid-chelated trace minerals into broiler diets as an alternative to industry levels of inorganic trace minerals has been found to significantly reduce FPD (Zhao et al., 2010; Manangi et al., 2012; Da Costa et al., 2016). Rations supplemented with additional fat (Fuhrmann and Kamphues, 2016), fatty acids (Khosravinia, 2015), vitamin D_3 (Sun et al., 2013), biotin (Abd El-Wahab et al., 2013; Sun et al., 2017), tannic acid (Cengiz et al., 2017) and a directly fed microbe (*Bacillus*) in combination with dietary enzymes (Dersjant-Li et al., 2015) have all been shown to reduce FPD. Dietary enzymes, such as phytase, are beneficial to foot health as they improve nutrient digestibility and reduce litter moisture content (Farhadi et al., 2017).

3.3.6 Lighting

Low-level lighting (0.5–5 lux) delivered with a short scotoperiod (period of darkness) appears to be a risk factor for contact dermatitis (Deep et al., 2013; Bassler et al., 2013; Schwean-Lardner et al., 2013; Senaratna et al., 2016) compared to birds reared under brighter lighting (10–320 lux). This is likely to be associated with an increase in body mass in the low-light birds (Lien et al., 2007; Blatchford et al., 2012; Senaratna et al., 2016). The type of lighting may also be important. The use of LED bulbs has been linked with improved FPD and HB scores when compared with compact fluorescent bulbs (Huth and Archer, 2015).

3.3.7 Egg quality, incubation and early life

At least some susceptibility to contact dermatitis appears to have its foundations in pre-hatch conditions. Lower HB scores have been reported for broilers originating from young parent breeders (particularly those individuals with a low hatching weight) than those originating from older parents (Henriksen et al., 2016).

Appropriate incubation temperatures also appear to be important. Oviedo-Rondón et al. (2009a) observed HB lesions to be more prevalent in 8-week-old commercial broilers hatched from multi-stage incubators (associated with greater temperature fluctuation) than those hatched from single-stage incubators. Fluctuating incubation temperatures were also seen to increase FPD susceptibility via alterations in foot skin structure (Da Costa et al., 2016). Broilers incubated at temperatures (39–40°C) higher than the optimum recommended range may also have a greater risk of developing severe HB in later life (Ipek and Sozcu, 2016).

Raising brooding temperature may help reduce the risk of HB development in heavy chicks. Henriksen et al. (2016) reported that a high brooding temperature (37°C) during the first week post-hatch increased activity, delayed body weight gain and reduced the prevalence of HB at 5 weeks in both high- and low-hatch-weight chicks (compared to high-hatch-weight chicks brooded at 33°C).

3.4 Welfare impact of contact dermatitis

Although there is no evidence for a relationship between contact dermatitis and mortality (de Jong et al., 2012a), skin lesions provide an obvious point of entry for pathogenic microorganisms, which can lead to secondary infections and advanced pathologies such as gangrenous dermatitis and osteomyelitis (Dinev, 2009). In broilers, both FPD and HB are associated with an increased prevalence and severity of *Campylobacter* infection (Bull et al., 2008; Rushton et al., 2009). Hepworth et al. (2011) reported HB to be a useful indicator of flock health since HB was positively associated with the percentage of birds with septicaemia and fever, detected at post-mortem inspection. In addition HB is positively associated with lameness (Sørensen et al., 2000; Haslam et al., 2007); presumably as broilers age and become heavier (and experience deteriorations in leg health), they spend more time lying down, increasing contact time with the litter.

De Jong et al. (2014) observed reductions in production parameters (including body weight) and walking ability (increased GS) in broilers reared on wet litter. As FPD also increased on wet litter the authors hypothesised that birds with deep or infected lesions may have experienced pain and consequently fed and drunk less due to inappetence (de Jong et al., 2014). Sherlock et al. (2012) compared global hepatic gene expression in control birds and in those with experimentally induced FPD and HB lesions. They reported evidence for the inflammatory reaction to impact upon key pathways linked with growth, metabolism and energy utilisation and hypothesised that pain may be the underlying trigger for the up-regulation of genes linked to a pro-inflammatory response and energy metabolism (Sherlock et al., 2012). Further evidence for FPD having a pain component is provided by Hothersall et al. (2016); birds with more severe FPD lesions sat sooner in an LTL test; however, standing ability was improved following NSAID treatment.

3.5 Prevention and control of contact dermatitis

3.5.1 Genetic selection

Since fast-growing genotypes have conclusively been shown to demonstrate greater susceptibility to developing contact dermatitis than the slower-growing genotypes (Sarica et al., 2014; Yamak et al., 2016), greater use of slow-growing strains in extended production systems would have direct and obvious benefits to broiler welfare. HB prevalence also varies according to the standard commercial hybrid used, and this diversity has been attributed to differences in both susceptibility and predisposition to generate wet litter (DEFRA, 2010). Breeding programmes have been shown to offer a direct means by which flock susceptibility to FPD and HB can be selected against and reduced without compromising further improvements in production parameters (Kjaer et al., 2006; Akbas et al., 2009; Ask, 2010; Kapell et al., 2012b), and such selection should continue.

3.5.2 Litter properties

The use of appropriate bedding material in floor-based poultry production systems is necessary to meet several health and welfare requirements. In regard to managing dermatitis deep litter systems should be 'deep', with at least a 7.6 cm depth of litter provided (Shepherd et al., 2017). Bedding material should have high absorbance properties (i.e. have a small particle size) as well as the ability to quickly release moisture (Bilgili et al., 2009; Cengiz et al., 2011). Flocks reared on wood-shavings or sawdust exhibit less FPD than those reared on chopped straw (e.g. Kyvsgaard et al., 2013; Skrbic et al., 2015; Villarroel et al., 2018), rice husks (e.g. Jacob et al., 2016b), grass (e.g. Garcia et al., 2012) or corncob litter (Xavier et al., 2010). Wood-shavings also appear to be the most appropriate litter for managing HB. Lower HB scores were observed at the end of the grow-out period in broilers reared on wood-shavings than paper sludge (Villagrá et al., 2011) or chopped straw (Skrbic et al., 2015). Unchopped straw is unsuitable for use as litter because it has low moisture absorbency and is associated with high FPD (Đukić Stojčić et al., 2016).

Although broiler chickens are commonly housed on sawdust and wood-shavings in some parts of Europe and North America, due to the lack of availability elsewhere, alternative litter sources are required in many parts of the world. Other bedding materials with a potential for maintaining healthy skin include pelleted straw (Berk, 2009; Avdalovic et al., 2017), sand (Simsek et al., 2009) and vermiculite (Yildiz et al., 2014). Similar levels of HB and FPD are reported for conventional wood-shaving litter systems and alternative perforated flooring (netting) systems (Li et al., 2017). The use of shredded paper, although cheap, readily available and highly absorbent, has proved to be unsuccessful in the control of contact dermatitis (Villagrá et al., 2011; Kheravii et al., 2017) since it becomes sodden and fails to release moisture.

In some countries, due to costs and availability, litter (at least the lower layer) is often re-used. Although some studies report a higher prevalence of contact dermatitis in broilers reared on re-used litter than in those reared on fresh (HB: Jacob et al., 2016b; FPD: de Oliveira et al., 2015; Shepherd et al., 2017), other studies report the opposite (Jacob et al., 2016b; Yamak et al., 2016). The ability to successfully re-use sawdust litter appears to be limited to production facilities located in hot dry climates, which exploit effective litter management, ventilation and composting systems.

Litter additives have been shown to be useful in improving litter quality, such as the addition of microorganisms and enzymes to increase the metabolisation of organic components. The application of a biological promoter aimed at wastewater treatment (Micropan®Simplex, Eurovix USA) to chopped straw successfully reduced both litter pH and FPD prevalence compared to a control flock (Đukić Stojčić et al., 2016; Zikic et al., 2017). The addition of sodium bisulphate complex (SBS) to litter has also been shown to be beneficial for foot pad health; this treatment also acidifies the litter, thereby reducing the microbial growth of nitrifying bacteria and NH_3 formation (Toppel et al., 2019).

3.5.3 Managing litter moisture

Litter moisture levels are directly linked to air temperature and RH within the broiler house, which are themselves directly influenced by ambient air temperature and season, the choice of heating system, the ventilation rate, the stocking density and broiler body mass. Deviations in environmental parameters outside of the optimum range recommended by the breeder companies have been identified as a key factor adversely influencing FPD (Jones et al., 2005). A detrimental effect of stocking density on RH has been observed to occur towards the end of the production cycle, indicating that environmental control becomes more difficult at this time (Jones et al., 2005). Presumably the ventilation rate requires to be increased once the birds get larger, less active and cover more of the litter surface, to improve air flow at the litter surface and remove litter moisture. With appropriate environmental management/control and the maintenance of temperature and RH levels within optimum limits, it should be possible to maintain good-quality dry litter throughout the production cycle.

Season has a marked influence upon contact dermatitis, with more lesions observed during colder months (Meluzzi et al., 2008a; Hepworth et al., 2010; Dinev et al., 2019). RH and litter moisture are also usually higher during colder months (e.g. Dawkins et al., 2004; Hermans et al., 2006; Meluzzi et al., 2008a). To save heating costs and maintain shed temperatures, ventilation rates are often reduced at this time. Gas-burning space heaters are a common form of heating in older UK broiler houses (Jones et al., 2005). These systems are located within the sheds themselves and generate a lot of water vapour as a by-product of combustion.

Gut health is linked with litter quality as any increase in faecal water content will be directly transferred to the litter. Causes of diarrhoea (scour) in broilers include coccidiosis, worms, viral or bacterial infection, a diet too high in protein, abrupt dietary alterations or long periods without feed (e.g. caused by feed equipment failures) and biosecurity breaches (Hermans et al., 2006). The use of antibiotics has been shown to improve broiler gut health, indirectly improve litter quality and reduce FPD severity (de Jong et al., 2012a). Global pressure to reduce antibiotic usage in all forms of livestock production has recently sparked much interest in the impact of intestinal microbiota upon broiler productivity (improved feed conversion efficiency) and health (e.g. Stanley et al., 2014; Oviedo-Rondón, 2019). Healthy gut microbiota lowers the risk of enteric diseases due to a reduction in pathogenic load (via competitive exclusion) and improved immune function. A reduction in enteric diseases should, in turn, improve litter quality. The use of probiotics, prebiotics and synbiotics (mixtures of both) to promote beneficial gut microbiota is emerging as a possible alternative to in-feed antibiotics. Results are very encouraging; some studies have even demonstrated prebiotic treatment to be more effective than antibiotics at maintaining broiler intestinal health (e.g. Al-Baadani et al., 2016).

The majority of broiler farms now use nipple and/or cup drinkers as opposed to bell drinkers since they minimise water spillage, and water spillage directly reduces litter quality. Increased water consumption has also been directly associated with FPD (Manning et al., 2007). Daily water consumption is currently monitored in l/bird/day to monitor flock performance and provide an indication of bird health. However, it has been shown that monitoring water consumption as l/m^2 floor area/day is a good lag (end of crop) indicator of litter quality (Manning et al., 2007). The use of automatic water meters has been linked with a decrease in HB (Hepworth et al., 2010). An ability to monitor and control water intake may, therefore, facilitate better stockmanship and improve litter management.

3.5.4 Environmental enrichment

Although evidence for a positive influence of elevated structures on contact dermatitis is variable and likely to reflect a different uptake in perch use between studies, the provision of straw bales and perches has been reported to reduce contact dermatitis (FPD: Ventura et al., 2010; Hongchao et al., 2013; Ohara et al., 2015; Kiyma et al., 2016; HB: Karaarslan and Nazligul, 2018). Elevated structures encourage better vertical distribution of birds within the available space, decreasing bird density at the floor level and allowing better circulation of air at the litter surface. They also provide a cleaner and dryer surface for the birds to sit and stand on, away from the litter surface.

3.5.5 Monitoring and surveillance systems

Camera systems are being developed to continuously model and monitor the movement and activity of broiler flocks throughout the production cycle to automatically detect and trigger alerts for FPD and HB (Dawkins et al., 2017; Fernandez et al., 2018). One system, in particular, shows great potential to inform improved flock management by providing farmers with an early warning of welfare issues. A camera-based optical flow system (OPTICFLOCK, see Chapter 7) has been shown to have a greater capacity to predict the prevalence of FPD and HB at slaughter (even before external signs become visible) than water consumption, bodyweight or mortality data (Dawkins et al., 2017).

FPD is an important welfare indicator. It is encouraging that government surveillance schemes, whereby producers have to reduce their stocking density or correct management deficiencies (e.g. improve litter quality) if their overall flock score (calculated from the prevalence and severity of FPD lesions at slaughter) exceeds certain predefined trigger limits, have proved to be successful at reducing FPD in Sweden, Denmark (e.g. Kyvsgaard et al., 2013) and the UK (Part et al., 2016). It is hoped that other countries will adopt similar national monitoring schemes and that a widespread uptake of automated monitoring systems will soon provide a means for continuous on-farm surveillance.

4 Conclusion and future trends

Broiler lameness and contact dermatitis remain commonplace within intensively reared commercial flocks, yet it is difficult to infer whether there have been significant improvements or indeed global declines in broiler welfare over recent years. Drawing any definitive conclusions is problematic since data remain sparse and are collected in different ways, multiple management differences exist both within and between countries and the breeds are undergoing constant change (via genetic selection). Leg problems certainly occur more frequently in heavy fast-growing strains due to muscular tissue accumulation occurring at a faster rate than the developing skeleton can support. Although lameness becomes more apparent at higher stocking densities (likely due to forced inactivity), environmental management, rather than stocking density, appears to be critical in the prevention of wet litter and contact dermatitis. Appropriate ventilation should be provided year-round, and both temperature and RH should remain within the recommended range at all times.

Broilers display low activity levels as a consequence of genetic selection for an unnatural body morphology and an excessive body mass; accordingly, they spend the majority of their time sitting in contact with the litter surface. The provision of elevated structures (such as straw bales and platforms with ramp-access), in combination with appropriate lighting, should facilitate rest and encourage activity, ultimately benefiting bone strength and leg and skin health. The provision of slow-growing genotypes with a suitable well-drained range should potentially encourage broilers to perform more activities outside, away from the litter surface of the barn, and thereby limit dermatitis progression.

Both lameness and contact dermatitis pose serious welfare concerns. Lameness physically limits behavioural expression and compromises mobility; birds may experience difficulty in gaining access to feeders and drinkers. Skin lesions provide a gateway for bacteria which can, in turn, trigger joint inflammation. Although evidence for an association between lameness, pain and underlying pathologies remains inconclusive, it is highly likely that some forms of lameness and severe footpad lesions are painful. More research is required to quantify the level of pain associated with specific pathologies, to enable us to differentiate these from other, morphological, gait abnormalities and to better focus breeding programmes (i.e. provide the genetics companies with targeted pressure).

Contact dermatitis and many leg pathologies are heritable and can be selected against. Although the use of genetic selection has had some success in the control of certain pathologies, the persistence of leg and skin conditions in commercial flocks indicates that more needs to be done. A greater uptake, commercially, of the slower-growing breeds would directly benefit broiler welfare. Non-governmental organisations (NGOs) have been, and will continue to be, fundamental in highlighting the requirement for robust genotypes. The Royal Society for Protection of Cruelty to Animals (RSPCA, UK) independently assesses new breeds according to a broiler breed welfare assessment protocol prior to its acceptance for use within the RSPCA-assured scheme; this includes key welfare outcome measures such as growth rate, leg heath and mortality. A number of animal protection organisations have recently set out an NGO 'Broiler Ask' initiative in the EU and the USA. The EU letter can be found at https://welfarecommitments.com/europeletter/. This scheme directly targets food suppliers and retailers to request that they raise welfare standards for meat chickens including better labelling on meat products, a maximum stocking density of 30 kg/m^2, the adoption of approved higher-welfare breeds and improved lighting by 2026.

Lameness, FPD and HB are generally assessed visually using severity scales, either on farm or at the processing plant. All three measures are used as flock welfare indicators and are crucial for setting (and reviewing) targets to progressively improve welfare within the industry. It is hoped that the development of fully automated video surveillance systems for use on farm will soon enable flock behaviour to be continuously monitored and any anomalies detected to be flagged up to provide producers with an early warning prior to the development of potential welfare issues.

Although standard broiler diets have been developed to deliver optimum nutrition, additional supplements have been observed to confer substantial health benefits. Dietary supplementation with prebiotics, probiotics and synbiotics has been shown to improve the gut uptake of beneficial nutrients and minerals, leading to improvements in the leg health. The addition of dietary enzymes improves nutrient digestibility and has been associated with improvements in both litter quality and footpad health. The uptake of increasingly integrated and automated management systems such as Flockman® (a combined lighting and feed control system for broilers) appears to have numerous welfare benefits. Such systems claim to improve digestion, feed conversion, litter quality and leg health via the automation of lighting programmes and provision of meal feeding to broiler flocks.

More needs to be done to help farmers recognise that the implementation of different feeding practices (e.g. whole wheat feeding, meal feeding and sequential feeding) could have a substantial positive impact upon both broiler welfare and farm economics.

The overarching problem preventing fundamental change within the poultry industry for the improvement of broiler skin and leg health is one of financial cost. The majority of factors that increase bird activity (i.e. appropriate environmental enrichment, lighting regimes, diet and gentotype) also improve skin and leg health; however, the producers (and the genetics companies) hesitate to make these changes due to a real or perceived association of increased activity with lowered production (food conversion) parameters. In fact, it is entirely possible that extra production costs resulting from the implementation of certain management practices designed to reduce lameness and/or improve skin health could be compensated for by other gains. For example, improving leg health and increasing flock uniformity have the potential to confer many economical benefits, including better technical performance (and thus reduced production costs), lowered mortality and reduced carcass condemnation rates. Comprehensive economic cost-benefit analyses may therefore be necessary to bolster both consumer pressure and ethical persuasion for change.

5 Where to look for further information

Further reading

- *Measuring and Auditing Broiler Welfare* by Weeks and Butterworth (2004) provides a good introduction to lameness and contact dermatitis.

- For a useful recent review of the applied ethology of broilers, including space use, stocking density, behaviour and welfare, see *The Behavioural Biology of Chickens* by Nicol (2015).
- *Poultry Feathers and Skin: The Poultry Integument in Health and Welfare*, edited by Olukosi et al. (2019), contains chapters on contact dermatitis and the genetics of contact dermatitis.
- Bradshaw, R. H., Kirkden, R. D. and Broom, D. M. 2002. A review of the aetiology and pathology of leg weakness in broilers in relation to their welfare. *Avian and Poultry Biological Reviews* 13(2), 45–103.
- Dinev, I. 2012. Leg weakness pathology in broiler chickens. *Journal of Poultry Science* 49(2), 63–67.
- Kierończyk, B., Rawski, M., Józefiak, D. and Świątkiewicz, S. 2017. Infectious and non-infectious factors associated with leg disorders in poultry – a review. *Annals of Animal Science* 17(3), 645–69.
- Mayne, R. 2005. A review of the aetiology and possible causative factors of foot pad dermatitis in growing turkeys and broilers. *World's Poultry Science Journal* 61(2), 256–67.
- Pedersen, I. J. and Forkman, B. 2019. Improving leg health in broiler chickens: a systematic review of the effect of environmental enrichment. *Animal Welfare* 28(2), 215–30.
- Swiatkiewicz, S., Arczewska-Wlosek, A. and Jozefiak, D. 2017. The nutrition of poultry as a factor affecting litter quality and foot pad dermatitis – an updated review. *Journal of Animal Physiology and Animal Nutrition (Berlin)* 101(5), e14–20. doi:10.1111/jpn.12630.
- Waldenstedt, L. 2006. Nutritional factors of importance for optimal leg health in broilers: a review. *Animal Feed Science and Technology* 126(3–4), 291–307.
- Wideman, R. F. 2016. Bacterial chondronecrosis with osteomyelitis and lameness in broilers: a review. *Poultry Science* 95(2), 325–44.

Key Journals

- *Animal Welfare*
- *British Poultry Science*
- *Journal of Applied Poultry Research*
- *Poultry Science*
- *Veterinary Record*

Key Conferences

- European Symposium on Poultry Welfare
- UFAW Animal Welfare Conference: Recent Advances in Animal Welfare Science

- WAFL: International Conference on the Assessment of Animal Welfare at Farm and Group Level

6 References

Abd El-Wahab, A., Radko, D. and Kamphues, J. 2013. High dietary levels of biotin and zinc to improve health of foot pads in broilers exposed experimentally to litter with critical moisture content. *Poultry Science* 92(7), 1774–82. doi:10.3382/ps.2013-03054.

Abd El-Wahab, A., Visscher, C. and Kamphues, J. 2018. Impact of different dietary protein sources on performance, litter quality and foot pad dermatitis in broilers. *Journal of Animal and Feed Sciences* 27(2), 148–54. doi:10.22358/jafs/90696/2018.

Abdulla, N. R., Loh, T. C., Akit, H., Sazili, A. Q., Foo, H. L., Kareem, K. Y., Mohamad, R. and Abdul Rahim, R. 2017. Effects of dietary oil sources, calcium and phosphorus levels on growth performance, carcass characteristics and bone quality of broiler chickens. *Journal of Applied Animal Research* 45(1), 423–9. doi:10.1080/09712119.2016.1206903.

Aguado, E., Pascaretti-Grizon, F., Goyenvalle, E., Audran, M. and Chappard, D. 2015. Bone mass and bone quality are altered by hypoactivity in the chicken. *PLoS ONE* 10(1), e0116763. doi:10.1371/journal.pone.0116763.

Akbaş, Y., Yalcin, S., Ozkan, S., Kirkpinar, F., Takma, C., Gevrekçi, Y,, Guller, H. C. and Turkmut, L. 2009. Heritability estimates of tibial dyschondroplasia, valgus-varus, foot-pad dermatitis and hock burn in broiler. *Archiv fur Geflugelkunde* 73(1), 1–6.

Al-Baadani, H. H., Abudabos, A. M., Al-Mufarrej, S. I. and Alzawqari, M. 2016. Effects of dietary inclusion of probiotics, prebiotics and Synbiotics on intestinal histological changes in challenged broiler chickens. *South African Journal of Animal Science* 46(2), 157–65. doi:10.4314/sajas.v46i2.6.

Allain, V., Mirabito, L., Arnould, C., Colas, M., Le Bouquin, S., Lupo, C. and Michel, V. 2009. Skin lesions in broiler chickens measured at the slaughterhouse: relationships between lesions and between their prevalence and rearing factors. *British Poultry Science* 50(4), 407–17. doi:10.1080/00071660903110901.

Al-Rubaye, A. A. K., Ekesi, N. S., Zaki, S., Emami, N. K., Wideman, R. F. and Rhoads, D. D. 2017. Chondronecrosis with osteomyelitis in broilers: further defining a bacterial challenge model using the wire flooring model. *Poultry Science* 96(2), 332–40. doi:10.3382/ps/pew299.

Alvino, G. M., Archer, G. S. and Mench, J. A. 2009. Behavioural time budgets of broiler chickens reared in varying light intensities. *Applied Animal Behaviour Science* 118(1–2), 54–61. doi:10.1016/j.applanim.2009.02.003.

Arnould, C. and Faure, J. M. 2004. Use of pen space and activity of broiler chickens reared at two different densities (vol 84, pg 281, 2003). *Applied Animal Behaviour Science* 87(1–2), 153–70.

Ask, B. 2010. Genetic variation of contact dermatitis in broilers. *Poultry Science* 89(5), 866–75. doi:10.3382/ps.2009-00496.

Avdalovic, V., Vueinic, M., Resanovic, R., Avdalovic, J., Maslic-Strizaks, D. and Vucicevic, M. 2017. Effect of pelleted and chopped wheat straw on the footpad dermatitis in broilers. *Pakistan Journal of Zoology* 49(5), 1639–46. doi:10.17582/journal.pjz/2017.49.5.1639.1646.

Aydin, A. 2017a. Development of an early detection system for lameness of broilers using computer vision. *Computers and Electronics in Agriculture* 136, 140–6. doi:10.1016/j.compag.2017.02.019.

Aydin, A. 2017b. Using 3D vision camera system to automatically assess the level of inactivity in broiler chickens. *Computers and Electronics in Agriculture* 135, 4–10. doi:10.1016/j.compag.2017.01.024.

Aydin, A., Bahr, C. and Berckmans, D. 2015. Automatic classification of measures of lying to assess the lameness of broilers. *Animal Welfare* 24(3), 335–43. doi:10.7120/09627286.24.3.335.

Bailie, C. L. and O'Connell, N. E. 2015. The influence of providing perches and string on activity levels, fearfulness and leg health in commercial broiler chickens. *Animal* 9(4), 660–8. doi:10.1017/S1751731114002821.

Bailie, C. L., Ball, M. E. E. and O'Connell, N. E. 2013. Influence of the provision of natural light and straw bales on activity levels and leg health in commercial broiler chickens. *Animal* 7(4), 618–26. doi:10.1017/S1751731112002108.

Bailie, C. L., Baxter, M. and O'Connell, N. E. 2018a. Exploring perch provision options for commercial broiler chickens. *Applied Animal Behaviour Science* 200, 114–22. doi:10.1016/j.applanim.2017.12.007.

Bailie, C. L., Ijichi, C. and O'Connell, N. E. 2018b. Effects of stocking density and string provision on welfare-related measures in commercial broiler chickens in windowed houses. *Poultry Science* 97(5), 1503–10. doi:10.3382/ps/pey026.

Baracho, M. S., Naas, I. A., Bueno, L. G. F., Nascimento, G. R. and Moura, D. J. 2012. Broiler walking ability and toe asymmetry under harsh rearing conditions. *Revista Brasileira de Ciência Avícola* 14(3), 217–22. doi:10.1590/S1516-635X2012000300009.

Bassler, A. W., Arnould, C., Butterworth, A., Colin, L., De Jong, I. C., Ferrante, V., Ferrari, P., Haslam, S., Wemelsfelder, F. and Blokhuis, H. J. 2013. Potential risk factors associated with contact dermatitis, lameness, negative emotional state, and fear of humans in broiler chicken flocks. *Poultry Science* 92(11), 2811–26. doi:10.3382/ps.2013-03208.

Baxter, M., Bailie, C. L. and O'Connell, N. E. 2018. Evalution of dustbathing substrate and straw bales as environmental enrichments in commercial broiler housing. *Applied Animal Behaviour Science* 200, 78–85. doi:10.1016/j.applanim.2017.11.010.

Bayram, A. and Özkan, S. 2010. Effects of a 16-hour light, 8-hour dark lighting schedule on behavioral traits and performance in male broiler chickens. *Journal of Applied Poultry Research* 19(3), 263–73. doi:10.3382/japr.2009-00026.

Berg, C. and Sanotra, G. S. 2003. Can a modified latency-to-lie test be used to validate gait-scoring results in commercial broiler flocks? *Animal Welfare* 12(4), 655–9.

Berk, J. 2009. Effect of litter type on prevalence and severity of pododermatitis in male broilers. *Berliner Und Munchener Tierarztliche Wochenschrift* 122(7–8), 257–63.

Bilgili, S. F., Hess, J. B., Blake, J. P., Macklin, K. S., Saenmahayak, B. and Sibley, J. L. 2009. Influence of bedding material on footpad dermatitis in broiler chickens. *Journal of Applied Poultry Research* 18(3), 583–9. doi:10.3382/japr.2009-00023.

Birgul, O. B., Mutaf, S. and Alkan, S. 2012. Effects of different angled perches on leg disorders in broilers. *Archiv fur Geflugelkunde* 76(1), 44–8.

Bizeray, D., Estevez, I., Leterrier, C. and Faure, J. M. 2002a. Effects of increasing environmental complexity on the physical activity of broiler chickens. *Applied Animal Behaviour Science* 79(1), 27–41. doi:10.1016/S0168-1591(02)00083-7.

Bizeray, D., Estevez, I., Leterrier, C. and Faure, J. M. 2002b. Influence of increased environmental complexity on leg condition, performance, and level of fearfulness in broilers. *Poultry Science* 81(6), 767–73. doi:10.1093/ps/81.6.767.

Blatchford, R. A., Klasing, K. C., Shivaprasad, H. L., Wakenell, P. S., Archer, G. S. and Mench, J. A. 2009. The effect of light intensity on the behavior, eye and leg health, and immune function of broiler chickens. *Poultry Science* 88(1), 20–8. doi:10.3382/ps.2008-00177.

Blatchford, R. A., Archer, G. S. and Mench, J. A. 2012. Contrast in light intensity, rather than day length, influences the behavior and health of broiler chickens. *Poultry Science* 91(8), 1768–74. doi:10.3382/ps.2011-02051.

Bokkers, E. A. M. and Koene, P. 2003. Behaviour of fast- and slow growing broilers to 12 weeks of age and the physical consequences. *Applied Animal Behaviour Science* 81(1), 59–72. doi:10.1016/S0168-1591(02)00251-4.

Bokkers, E. A. M. and Koene, P. 2004. Motivation and ability to walk for a food reward in fast- and slow-growing broilers to 12 weeks of age. *Behavioural Processes* 67(2), 121–30. doi:10.1016/j.beproc.2004.03.015.

Bokkers, E. A. M., Zimmerman, P. H., Rodenburg, T. B. and Koene, P. 2007. Walking behaviour of heavy and light broilers in an operant runway test with varying durations of feed deprivation and feed access. *Applied Animal Behaviour Science* 108(1–2), 129–42. doi:10.1016/j.applanim.2006.10.011.

Bradbury, E. J., Wilkinson, S. J., Cronin, G. M., Thomson, P. C., Bedford, M. R. and Cowieson, A. J. 2014. Nutritional geometry of calcium and phosphorus nutrition in broiler chicks. Growth performance, skeletal health and intake arrays. *Animal* 8(7), 1071–9. doi:10.1017/S1751731114001037.

Bradbury, E. J., Wilkinson, S. J., Cronin, G. M., Thomson, P., Walk, C. L. and Cowieson, A. J. 2017. Evaluation of the effect of a highly soluble calcium source in broiler diets supplemented with phytase on performance, nutrient digestibility, foot ash, mobility and leg weakness. *Animal Production Science* 57(10), 2016–26. doi:10.1071/AN16142.

Bradshaw, R. H., Kirkden, R. D. and Broom, D. M. 2002. A review of the aetiology and pathology of leg weakness in broilers in relation to welfare. *Avian and Poultry Biology Reviews* 13(2), 45–103. doi:10.3184/147020602783698421.

Brickett, K. E., Dahiya, J. P., Classen, H. L., Annett, C. B. and Gomis, S. 2007. The impact of nutrient density, feed form, and photoperiod on the walking ability and skeletal quality of broiler chickens. *Poultry Science* 86(10), 2117–25. doi:10.1093/ps/86.10.2117.

Broom, D. M. and Reefmann, N. 2005. Chicken welfare as indicated by lesions on carcases in supermarkets. *British Poultry Science* 46(4), 407–14. doi:10.1080/00071660500181149.

Buijs, S., Keeling, L., Rettenbacher, S., Van Poucke, E. and Tuyttens, F. A. M. 2009. Stocking density effects on broiler welfare: identifying sensitive ranges for different indicators. *Poultry Science* 88(8), 1536–43. doi:10.3382/ps.2009-00007.

Buijs, S., Keeling, L. J., Vangestel, C., Baert, J. and Tuyttens, F. A. M. 2011. Neighbourhood analysis as an indicator of spatial requirements of broiler chickens. *Applied Animal Behaviour Science* 129(2–4), 111–20. doi:10.1016/j.applanim.2010.11.017.

Buijs, S., Van Poucke, E., Van Dongen, S., Lens, L., Baert, J. and Tuyttens, F. A. 2012. The influence of stocking density on broiler chicken bone quality and fluctuating asymmetry. *Poultry Science* 91(8), 1759–67. doi:10.3382/ps.2011-01859.

Bull, S. A., Thomas, A., Humphrey, T., Ellis-Iversen, J., Cook, A. J., Lovell, R. and Jorgensen, F. 2008. Flock health indicators and *Campylobacter* spp. in commercial housed broilers reared in Great Britain. *Applied and Environmental Microbiology* 74(17), 5408–13. doi:10.1128/AEM.00462-08.

Butterworth, A. and Haslam, S. M. 2009. A lameness control strategy for broiler fowls. Welfare Quality Reports No. 13. Cardiff School of City and Regional Planning, Cardiff, UK. Available at: http://www.welfarequality.net/media/1119/wqr13.pdf.

Butterworth, A., Weeks, C. A., Crea, P. R. and Kestin, S. C. 2002. Dehydration and lameness in a broiler flock. *Animal Welfare* 11(1), 89–94.

Caplen, G., Hothersall, B., Murrell, J. C., Nicol, C. J., Waterman-Pearson, A. E., Weeks, C. A. and Colborne, G. R. 2012. Kinematic analysis quantifies gait abnormalities associated with lameness in broiler chickens and identifies evolutionary gait differences. *PLoS ONE* 7(7), e40800. doi:10.1371/journal.pone.0040800.

Caplen, G., Colborne, G. R., Hothersall, B., Nicol, C. J., Waterman-Pearson, A. E., Weeks, C. A. and Murrell, J. C. 2013a. Lame broiler chickens respond to non-steroidal anti-inflammatory drugs with objective changes in gait function: a controlled clinical trial. *Veterinary Journal* 196(3), 477–82. doi:10.1016/j.tvjl.2012.12.007.

Caplen, G., Baker, L., Hothersall, B., McKeegan, D. E. F., Sandilands, V., Sparks, N. H., Waterman-Pearson, A. E. and Murrell, J. C. 2013b. Thermal nociception as a measure of non-steroidal anti-inflammatory drug effectiveness in broiler chickens with articular pain. *Veterinary Journal* 198(3), 616–9. doi:10.1016/j.tvjl.2013.09.013.

Caplen, G., Hothersall, B., Nicol, C. J., Parker, R. M. A., Waterman-Pearson, A. E., Weeks, C. and Murrell, J. 2014. Lameness is consistently better at predicting broiler chicken performance in mobility tests than other broiler characteristics. *Animal Welfare* 23(2), 179–87. doi:10.7120/09627286.23.2.179.

Castellini, C., Mugnai, C., Moscati, L., Mattioli, S., Amato, M. G., Cartoni Mancinelli, A. and Dal Bosco, A. 2016. Adaptation to organic rearing system of eight different chicken genotypes: behaviour, welfare and performance. *Italian Journal of Animal Science* 15(1), 37–46. doi:10.1080/1828051X.2015.1131893.

Cengiz, Ö, Hess, J. B. and Bilgili, S. F. 2011. Effect of bedding type and transient wetness on footpad dermatitis in broiler chickens. *Journal of Applied Poultry Research* 20(4), 554-60. doi:10.3382/japr.2011-00368.

Cengiz, O., Hess, J. B. and Bilgili, S. F. 2013. Effect of protein source on the development of footpad dermatitis in broiler chickens reared on different flooring types. *Archiv fur Geflugelkunde* 77(3), 166–70.

Cengiz, Ö, Koksal, B. H., Tatli, O., Sevim, Ö, Ahsan, U., Bilgili, S. F. and Gökhan Önol, A. 2017. Effect of dietary tannic acid supplementation in corn- or barley-based diets on growth performance, intestinal viscosity, litter quality, and incidence and severity of footpad dermatitis in broiler chickens. *Livestock Science* 202, 52–7. doi:10.1016/j.livsci.2017.05.016.

Charles, J. F., Ermann, J. and Aliprantis, A. O. 2015. The intestinal microbiome and skeletal fitness: connecting bugs and bones. *Clinical Immunology* 159(2), 163–9. doi:10.1016/j.clim.2015.03.019.

Classen, H. L. 2004. Day length affects performance, health and condemnations in broiler chickens. *Proceedings of the 16th Australian Poultry Science Symposium*, Sydney, New South Wales, Australia, 9–11 February 2004, pp. 11215. Available at: https://sydney.edu.au/vetscience/apss/documents/2004/APSS2004-classen-pp112-115.pdf.

Collins, L. M. 2008. Non-intrusive tracking of commercial broiler chickens *in situ* at different stocking densities. *Applied Animal Behaviour Science* 112(1-2), 94-105. doi:10.1016/j.applanim.2007.08.009.

Corr, S. A., Gentle, M. J., McCorquodale, C. C. and Bennett, D. 2003a. The effect of morphology on walking ability in the modern broiler: a gait analysis study. *Animal Welfare* 12(2), 159-71.

Corr, S. A., Gentle, M. J., McCorquodale, C. C. and Bennett, D. 2003b. The effect of morphology on the musculoskeletal system of the modern broiler. *Animal Welfare* 12(2), 145-57.

Corr, S. A., Maxwell, M., Gentle, M. J. and Bennett, D. 2003c. Preliminary study of joint disease in poultry by the analysis of synovial fluid. *Veterinary Record* 152(18), 549-54. doi:10.1136/vr.152.18.549.

Corr, S. A., McCorquodale, C., McDonald, J., Gentle, M. and McGovern, R. 2007. A force plate study of avian gait. *Journal of Biomechanics* 40(9), 2037-43. doi:10.1016/j.jbiomech.2006.09.014.

Coto, C., Yan, F., Cerrate, S., Wang, Z., Sacakli, P., Halley, J. T., Wiernusz, C. J., Martinez, A. and Waldroup, P. W. 2008. Effects of dietary levels of calcium and nonphytate P in broiler starter diets on live performance, bone development, and growth plate conditions in male broilers fed a corn-based diet. *International Journal of Poultry Science* 7(7), 638-45. doi:10.3923/ijps.2008.638.645.

Da Costa, M. J., Oviedo-Rondon, E. O., Wineland, M. J., Claassen, K. and Osborne, J. 2016. Effects of incubation temperatures and trace mineral sources on chicken live performance and footpad skin development. *Poultry Science* 95(4), 749-59. doi:10.3382/ps/pev446.

Dal Bosco, A. D., Mugnai, C., Amato, M. G., Piottoli, L., Cartoni, A. and Castellini, C. 2014. Effect of slaughtering age in different commercial chicken genotypes reared according to the organic system: 1. Welfare, carcass and meat traits. *Italian Journal of Animal Science* 13(2), 3308. doi:10.4081/ijas.2014.3308.

Danbury, T. C., Weeks, C. A., Chambers, J. P., Waterman-Pearson, A. E. and Kestin, S. C. 2000. Self-selection of the analgesic drug carprofen by lame broiler chickens. *Veterinary Record* 146(11), 307-11. doi:10.1136/vr.146.11.307.

Dantzer, R. 2009. Cytokine, sickness behavior, and depression. *Immunology and Allergy Clinics of North America* 29(2), 247-64. doi:10.1016/j.iac.2009.02.002.

Das, H. and Lacin, E. 2014. The effect of different photoperiods and stocking densities on fattening performance, carcass and some stress parameters in broilers. *Israel Journal of Veterinary Medicine* 69(4), 211-20.

Dawkins, M. S., Donnelly, C. A. and Jones, T. A. 2004. Chicken welfare is influenced more by housing conditions than by stocking density. *Nature* 427(6972), 342-4. doi:10.1038/nature02226.

Dawkins, M. S., Lee, H. J., Waitt, C. D. and Roberts, S. J. 2009. Optical flow patterns in broiler chicken flocks as automated measures of behaviour and gait. *Applied Animal Behaviour Science* 119(3-4), 203-9. doi:10.1016/j.applanim.2009.04.009.

Dawkins, M. S., Cain, R., Merelie, K. and Roberts, S. J. 2013. In search of the behavioural correlates of optical flow patterns in the automated assessment of broiler chicken welfare. *Applied Animal Behaviour Science* 145(1-2), 44-50. doi:10.1016/j.applanim.2013.02.001.

Dawkins, M. S., Roberts, S. J., Cain, R. J., Nickson, T. and Donnelly, C. A. 2017. Early warning of footpad dermatitis and hockburn in broiler chicken flocks using optical flow, bodyweight and water consumption. *Veterinary Record* 180(20), 499. doi:10.1136/vr.104066.

de Jong, I. C. and Gunnink, H. 2019. Effects of a commercial broiler enrichment programme with or without natural light on behaviour and other welfare indicators. *Animal* 13(2), 384–91. doi:10.1017/S1751731118001805.

de Jong, I. C., van Ham, J., Gunnink, H., Hindle, V. A. and Lourens, A. 2012a. Footpad dermatitis in Dutch broiler flocks: prevalence and factors of influence. *Poultry Science* 91(7), 1569–74. doi:10.3382/ps.2012-02156.

de Jong, I. C., van Harn, J., Gunnink, H., Lourens, A. and van Riel, J. W. 2012b. Measuring foot-pad lesions in commercial broiler houses. Some aspects of methodology. *Animal Welfare* 21(3), 325–30. doi:10.7120/09627286.21.3.325.

de Jong, I. C., Gunnink, H. and van Harn, J. 2014. Wet litter not only induces footpad dermatitis but also reduces overall welfare, technical performance, and carcass yield in broiler chickens. *Journal of Applied Poultry Research* 23(1), 51–8. doi:10.3382/japr.2013-00803.

de Oliveira, M. C., Goncalves, B. N., Padua, G. T., da Silva, V. G., da Silva, D. V. and Freitas, A. I. M. 2015. Treatment of poultry litter does not improve performance or carcass lesions in broilers. *Revista Colombiana De Ciencias Pecuarias* 28(4), 331–8.

Deep, A., Raginski, C., Schwean-Lardner, K., Fancher, B. I. and Classen, H. L. 2013. Minimum light intensity threshold to prevent negative effects on broiler production and welfare. *British Poultry Science* 54(6), 686–94. doi:10.1080/00071668.2013.847526.

Department of the Environment Food and Rural Affairs. 2010. Foot pad dermatitis & hock burn in broilers: risk factors, aetiology & welfare consequences. London, UK. Available at: http://sciencesearch.defra.gov.uk/Document.aspx?Document=AW1137SID5FinalReport.pdf (accessed on 16 June 2019).

Dersjant-Li, Y., van de Belt, K., van der Klis, J. D., Kettunen, H., Rinttila, T. and Awati, A. 2015. Effect of multi-enzymes in combination with a direct-fed microbial on performance and welfare parameters in broilers under commercial production settings. *Journal of Applied Poultry Research* 24(1), 80–90. doi:10.3382/japr/pfv003.

Dinev, I. 2009. Clinical and morphological investigations on the prevalence of lameness associated with femoral head necrosis in broilers. *British Poultry Science* 50(3), 284–90. doi:10.1080/00071660902942783.

Dinev, I. and Kanakov, D. 2011. Deep pectoral myopathy: prevalence in 7 weeks old broiler chickens in Bulgaria. *Revue De Medecine Veterinaire* 162(6), 279–83.

Dinev, I., Denev, S. A. and Edens, F. W. 2012. Comparative clinical and morphological studies on the incidence of tibial dyschondroplasia as a cause of lameness in three commercial lines of broiler chickens. *Journal of Applied Poultry Research* 21(3), 637–44. doi:10.3382/japr.2010-00303.

Dinev, I., Denev, S., Vashin, I., Kanakov, D. and Rusenova, N. 2019. Pathomorphological investigations on the prevalence of contact dermatitis lesions in broiler chickens. *Journal of Applied Animal Research* 47(1), 129–34. doi:10.1080/09712119.2019.1584105.

Dozier, W. A., Thaxton, J. P., Purswell, J. L., Olanrewaju, H. A., Branton, S. L. and Roush, W. B. 2006. Stocking density effects on male broilers grown to 1.8 kilograms of body weight. *Poultry Science* 85(2), 344–51. doi:10.1093/ps/85.2.344.

Duff, S. R. I. and Thorp, B. H. 1985. Abnormal angulation/torsion of the pelvic appendicular skeleton in broiler fowl: morphological and radiological findings. *Research in Veterinary Science* 39(3), 313–9. doi:10.1016/S0034-5288(18)31720-X.

Đukić Stojčić, M., Bjedov, S., Zikic, D., Peric, L. and Milosevic, N. 2016. Effect of straw size and microbial amendment of litter on certain litter quality parameters, ammonia emission, and footpad dermatitis in broilers. *Archives Animal Breeding* 59(1), 131-7. doi:10.5194/aab-59-131-2016.

Eichner, G., Vieira, S. L., Torres, C. A., Coneglian, J. L. B., Freitas, D. M. and Oyarzabal, O. A. 2007. Litter moisture and footpad dermatitis as affected by diets formulated on an all-vegetable basis or having the inclusion of poultry by-product. *Journal of Applied Poultry Research* 16(3), 344-50. doi:10.1093/japr/16.3.344.

Eriksson, M., Waldenstedt, L., Elwinger, K., Engstrom, B. and Fossum, O. 2010. Behaviour, production and health of organically reared fast-growing broilers fed low crude protein diets including different amino acid contents at start. *Acta Agriculturae Scandinavica, Section A - Animal Science* 60(2), 112-24. doi:10.1080/09064702.2010.502243.

Estevez, I. 2007. Density allowances for broilers: where to set the limits? *Poultry Science* 86(6), 1265-72. doi:10.1093/ps/86.6.1265.

Estevez, I., Tablante, N., Pettit-Riley, R. L. and Carr, L. 2002. Use of cool perches by broiler chickens. *Poultry Science* 81(1), 62-9. doi:10.1093/ps/81.1.62.

FAO. 2010. Poultry welfare in developing countries - welfare issues in commercial broiler production, pp. 117-18. Available at: http://www.fao.org/3/a-al723e.pdf.

Farhadi, D., Karimi, A., Sadeghi, G., Rostamzadeh, J. and Bedford, M. R. 2017. Effects of a high dose of microbial phytase and myo-inositol supplementation on growth performance, tibia mineralization, nutrient digestibility, litter moisture content, and foot problems in broiler chickens fed phosphorus-deficient diets. *Poultry Science* 96(10), 3664-75. doi:10.3382/ps/pex186.

Fernandes, B. C. D., Martins, M. R. F. B., Mendes, A. A., Paz, I., Komiyama, C. M., Milbradt, E. L. and Martins, B. B. 2012. Locomotion problems of broiler chickens and its relationship with the gait score. *Revista Brasileira de Zootecnia - Brazilian Journal of Animal Science* 41(8), 1951-5. doi:10.1590/S1516-35982012000800021.

Fernandez, A. P., Norton, T., Tullo, E., van Hertem, T., Youssef, A., Exadaktylos, V., Vranken, E., Guarino, M. and Berckmans, D. 2018. Real-time monitoring of broiler flock's welfare status using camera-based technology. *Biosystems Engineering* 173, 103-14. doi:10.1016/j.biosystemseng.2018.05.008.

Fidan, E. D., Nazligul, A., Turkyilmaz, M. K., Aypak, S. Ü., Kilimci, F. S., Karaarslan, S. and Kaya, M. 2017. Effect of photoperiod length and light intensity on some welfare criteria, carcass, and meat quality characteristics in broilers. *Revista Brasileira de Zootecnia - Brazilian Journal of Animal Science* 46(3), 202-10. doi:10.1590/s1806-92902017000300004.

Fleming, R. H. 2008. Nutritional factors affecting poultry bone health. *Proceedings of the Nutrition Society* 67(2), 177-83. doi:10.1017/S0029665108007015.

Fuhrmann, R. and Kamphues, J. 2016. Effects of fat content and source as well as of calcium and potassium content in the diet on fat excretion and saponification, litter quality and foot pad health in broilers. *European Poultry Science* 80, 118. doi:10.1399/eps.2016.118.

Garcia, R., Almeida Paz, I., Caldara, F., Nääs, I., Bueno, L., Freitas, L., Graciano, J. and Sim, S. 2012. Litter materials and the incidence of carcass lesions in broilers chickens. *Revista Brasileira de Ciência Avícola* 14(1), 27-32. doi:10.1590/S1516-635X2012000100005.

Garner, J. P., Falcone, C., Wakenell, P., Martin, M. and Mench, J. A. 2002. Reliability and validity of a modified gait scoring system and its use in assessing tibial dyschondroplasia in broilers. *British Poultry Science* 43(3), 355-63. doi:10.1080/00071660120103620.

Gentle, M. J., Tilston, V. and McKeegan, D. E. F. 2001. Mechanothermal nociceptors in the scaly skin of the chicken leg. *Neuroscience* 106(3), 643–52. doi:10.1016/s0306-4522(01)00318-9.

Gocsik, É., Silvera, A. M., Hansson, H., Saatkamp, H. W. and Blokhuis, H. J. 2017. Exploring the economic potential of reducing broiler lameness. *British Poultry Science* 58(4), 337–47. doi:10.1080/00071668.2017.1304530.

Gouveia, K. G., Vaz-Pires, P. and da Costa, P. M. 2009. Welfare assessment of broilers through examination of haematomas, foot-pad dermatitis, scratches and breast blisters at processing. *Animal Welfare* 18(1), 43–8.

Granquist, E. G., Vasdal, G., de Jong, I. C. and Moe, R. O. 2019. Lameness and its relationship with health and production measures in broiler chickens. *Animal* 13(10), 2365–72. doi:10.1017/S1751731119000466.

Groves, P. J. and Muir, W. I. 2014. A meta-analysis of experiments linking incubation conditions with subsequent leg weakness in broiler chickens. *PLoS ONE* 9(7), e102682. doi:10.1371/journal.pone.0102682.

Groves, P. J. and Muir, W. I. 2017. Earlier hatching time predisposes Cobb broiler chickens to tibial dyschondroplasia. *Animal* 11(1), 112–20. doi:10.1017/S1751731116001105.

Guo, R. X., Li, Z. Y., Zhou, X. H., Huang, C. Y., Hu, Y. C., Geng, S., Chen, X., Li, Q., Pan, Z. and Jiao, X. 2019. Induction of arthritis in chickens by infection with novel virulent *Salmonella pullorum* strains. *Veterinary Microbiology* 228, 165–72. doi:10.1016/j.vetmic.2018.11.032.

Hashimoto, S., Yamazaki, K., Obi, T. and Takase, K. 2011. Footpad dermatitis in broiler chickens in Japan. *Journal of Veterinary Medical Science* 73(3), 293–7. doi:10.1292/jvms.10-0329.

Hashimoto, S., Yamazaki, K., Obi, T. and Takase, K. 2013. Relationship between severity of footpad dermatitis and carcass performance in broiler chickens. *Journal of Veterinary Medical Science* 75(11), 1547–9. doi:10.1292/jvms.13-0031.

Haslam, S. M., Brown, S. N., Wilkins, L. J., Kestin, S. C., Warriss, P. D. and Nicol, C. J. 2006. Preliminary study to examine the utility of using foot burn or hock burn to assess aspects of housing conditions for broiler chicken. *British Poultry Science* 47(1), 13–8. doi:10.1080/00071660500475046.

Haslam, S. M., Knowles, T. G., Brown, S. N., Wilkins, L. J., Kestin, S. C., Warriss, P. D. and Nicol, C. J. 2007. Factors affecting the prevalence of foot pad dermatitis, hock burn and breast burn in broiler chicken. *British Poultry Science* 48(3), 264–75. doi:10.1080/00071660701371341.

Henriksen, S., Bilde, T. and Riber, A. B. 2016. Effects of post-hatch brooding temperature on broiler behavior, welfare, and growth. *Poultry Science* 95(10), 2235–43. doi:10.3382/ps/pew224.

Hepworth, P. J., Nefedov, A. V., Muchnik, I. B. and Morgan, K. L. 2010. Early warning indicators for hock burn in broiler flocks. *Avian Pathology* 39(5), 405–9. doi:10.1080/03079457.2010.510500.

Hepworth, P. J., Nefedov, A. V., Muchnik, I. B. and Morgan, K. L. 2011. Hock burn: an indicator of broiler flock health. *Veterinary Record* 168(11), 303. doi:10.1136/vr.c6897.

Hermans, P. G., Fradkin, D., Muchnik, I. B. and Morgan, K. L. 2006. Prevalence of wet litter and the associated risk factors in broiler flocks in the United Kingdom. *Veterinary Record* 158(18), 615–22. doi:10.1136/vr.158.18.615.

Hester, P. Y., Enneking, S. A., Haley, B. K., Cheng, H. W., Einstein, M. E. and Rubin, D. A. 2013. The effect of perch availability during pullet rearing and egg laying on musculoskeletal health of caged White Leghorn hens. *Poultry Science* 92(8), 1972-80. doi:10.3382/ps.2013-03008.

Hongchao, J., Jiang, Y., Song, Z., Zhao, J., Wang, X. and Lin, H. 2013. Effect of perch type and stocking density on the behaviour and growth of broilers. *Animal Production Science* 54(7), 930-41. doi:10.1071/AN13184.

Hothersall, B., Caplen, G., Parker, R. M. A., Nicol, C. J., Waterman-Pearson, A. E., Weeks, C. A. and Murrell, J. C. 2014. Thermal nociceptive threshold testing detects altered sensory processing in broiler chickens with spontaneous lameness. *PLoS ONE* 9(5), e97883. doi:10.1371/journal.pone.0097883.

Hothersall, B., Caplen, G., Parker, R. M. A., Nicol, C. J., Waterman-Pearson, A. E., Weeks, C. and Murrell, J. 2016. Effects of carprofen, meloxicam and butorphanol on broiler chickens' performance in mobility tests. *Animal Welfare* 25(1), 55-67. doi:10.7120/09627286.25.1.055.

Hunter, J. M., Anders, S. A., Crowe, T., Korver, D. R. and Bench, C. J. 2017. Practical assessment and management of foot pad dermatitis in commercial broiler chickens: a field study. *Journal of Applied Poultry Research* 26(4), 593-604. doi:10.3382/japr/pfx019.

Huth, J. C. and Archer, G. S. 2015. Comparison of two LED light bulbs to a dimmable CFL and their effects on broiler chicken growth, stress, and fear. *Poultry Science* 94(9), 2027-36. doi:10.3382/ps/pev215.

Ipek, A. and Sozcu, A. 2016. The effects of eggshell temperature fluctuations during incubation on welfare status and gait score of broilers. *Poultry Science* 95(6), 1296-303. doi:10.3382/ps/pew056.

Ipek, A. and Sozcu, A. 2017. The effects of access to pasture on growth performance, behavioural patterns, some blood parameters and carcass yield of a slow-growing broiler genotype. *Journal of Applied Animal Research* 45(1), 464-9. doi:10.1080/09712119.2016.1214136.

Jacob, F. G., Baracho, M. S., Naas, I. A., Salgado, D. A. and Souza, R. 2016a. Incidence of pododermatitis in broiler reared under two types of environment. *Revista Brasileira de Ciência Avícola* 18(2), 247-54. doi:10.1590/1806-9061-2015-0047.

Jacob, F. G., Baracho, M. S., Naas, I. A., Lima, N. S. D., Salgado, D. D. and Souza, R. 2016b. Risk of incidence of hock burn and pododermatitis in broilers reared under commercial conditions. *Revista Brasileira de Ciência Avícola* 18(3), 357-62. doi:10.1590/1806-9061-2015-0183.

James, C., Asher, L., Herborn, K. and Wiseman, J. 2018. The effect of supplementary ultraviolet wavelengths on broiler chicken welfare indicators. *Applied Animal Behaviour Science* 209, 55-64. doi:10.1016/j.applanim.2018.10.002.

Jones, T. A., Donnelly, C. A. and Dawkins, M. S. 2005. Environmental and management factors affecting the welfare of chickens on commercial farms in the United Kingdom and Denmark stocked at five densities. *Poultry Science* 84(8), 1155-65. doi:10.1093/ps/84.8.1155.

Jordan, D., Stuhec, I. and Bessei, W. 2011. Effect of whole wheat and feed pellets distribution in the litter on broilers' activity and performance. *Archiv fur Geflugelkunde* 75(2), 98-103.

Julian, R. J. 1998. Rapid growth problems: ascites and skeletal deformities in broilers. *Poultry Science* 77(12), 1773-80. doi:10.1093/ps/77.12.1773.

Kapell, D. N., Hill, W. G., Neeteson, A. M., McAdam, J., Koerhuis, A. N. M. and Avendaño, S. 2012a. Twenty-five years of selection for improved leg health in purebred broiler lines and underlying genetic parameters. *Poultry Science* 91(12), 3032-43. doi:10.3382/ps.2012-02578.

Kapell, D. N., Hill, W. G., Neeteson, A. M., McAdam, J., Koerhuis, A. N. M. and Avendaño, S. 2012b. Genetic parameters of foot-pad dermatitis and body weight in purebred broiler lines in 2 contrasting environments. *Poultry Science* 91(3), 565-74. doi:10.3382/ps.2011-01934.

Karaarslan, S. and Nazligul, A. 2018. Effects of lighting, stocking density, and access to perches on leg health variables as welfare indicators in broiler chickens. *Livestock Science* 218, 31-6. doi:10.1016/j.livsci.2018.10.008.

Kaukonen, E., Norring, M. and Valros, A. 2017. Perches and elevated platforms in commercial broiler farms: use and effect on walking ability, incidence of tibial dyschondroplasia and bone mineral content. *Animal* 11(5), 864-71. doi:10.1017/S1751731116002160.

Kestin, S. C., Knowles, T. G., Tinch, A. E. and Gregory, N. G. 1992. Prevalence of leg weakness in broiler-chickens and its relationship with genotype. *Veterinary Record* 131(9), 190-4. doi:10.1136/vr.131.9.190.

Kestin, S. C., Gordon, S., Su, G. and Sorensen, P. 2001. Relationships in broiler chickens between lameness, liveweight, growth rate and age. *Veterinary Record* 148(7), 195-7. doi:10.1136/vr.148.7.195.

Kheravii, S. K., Swick, R. A., Choct, M. and Wu, S. B. 2017. Potential of pelleted wheat straw as an alternative bedding material for broilers. *Poultry Science* 96(6), 1641-7. doi:10.3382/ps/pew473.

Khosravinia, H. 2015. Effect of dietary supplementation of medium-chain fatty acids on growth performance and prevalence of carcass defects in broiler chickens raised in different stocking densities. *Journal of Applied Poultry Research* 24(1), 1-9. doi:10.3382/japr/pfu001.

Kittelsen, K. E., David, B., Moe, R. O., Poulsen, H. D., Young, J. F. and Granquist, E. G. 2017. Associations among gait score, production data, abattoir registrations, and postmortem tibia measurements in broiler chickens. *Poultry Science* 96(5), 1033-40. doi:10.3382/ps/pew433.

Kiyma, Z., Kucukyilmaz, K. and Orojpour, A. 2016. Effects of perch availability on performance, carcass characteristics, and footpad lesions in broilers. *Archives Animal Breeding* 59(1), 19-25. doi:10.5194/aab-59-19-2016.

Kjaer, J. B., Su, G., Nielsen, B. L. and Sorensen, P. 2006. Foot pad dermatitis and hock burn in broiler chickens and degree of inheritance. *Poultry Science* 85(8), 1342-8. doi:10.1093/ps/85.8.1342.

Knowles, T. G., Kestin, S. C., Haslam, S. M., Brown, S. N., Green, L. E., Butterworth, A., Pope, S. J., Pfeiffer, D. and Nicol, C. J. 2008. Leg disorders in broiler chickens: prevalence, risk factors and prevention. *PLoS ONE* 3(2), e1545. doi:10.1371/journal.pone.0001545.

Kwiatkowska, K., Winiarska-Mieczan, A. and Kwiecien, M. 2017. Feed additives regulating calcium homeostasis in the bones of poultry - a review. *Annals of Animal Science* 17(2), 303-16. doi:10.1515/aoas-2016-0031.

Kyvsgaard, N. C., Jensen, H. B., Ambrosen, T. and Toft, N. 2013. Temporal changes and risk factors for foot-pad dermatitis in Danish broilers. *Poultry Science* 92(1), 26-32. doi:10.3382/ps.2012-02433.

Leterrier, C., Vallee, C., Constantin, P., Chagneau, A. M., Lessire, M., Lescoat, P., Berri, C., Baéza, E., Bizeray, D. and Bouvarel, I. 2008. Sequential feeding with variations in energy and protein levels improves gait score in meat-type chickens. *Animal* 2(11), 1658–65. doi:10.1017/S1751731108002875.

Lewis, P. D., Danisman, R. and Gous, R. M. 2009. Photoperiodic responses of broilers. III. Tibial breaking strength and ash content. *British Poultry Science* 50(6), 673–9. doi:10.1080/00071660903365612.

Li, H., Wen, X., Alphin, R., Zhu, Z. and Zhou, Z. 2017. Effects of two different broiler flooring systems on production performances, welfare, and environment under commercial production conditions. *Poultry Science* 96(5), 1108–19. doi:10.3382/ps/pew440.

Lien, R. J., Hess, J. B., McKee, S. R., Bilgili, S. F. and Townsend, J. C. 2007. Effect of light intensity and photoperiod on live performance, heterophil-to-lymphocyte ratio, and processing yields of broilers. *Poultry Science* 86(7), 1287–93. doi:10.1093/ps/86.7.1287.

Louton, H., Bergmann, S., Reese, S., Erhard, M., Bachmeier, J., Rösler, B. and Rauch, E. 2018. Animal- and management-based welfare indicators for a conventional broiler strain in 2 barn types (Louisiana barn and closed barn). *Poultry Science* 97(8), 2754–67. doi:10.3382/ps/pey111.

Lund, V. P., Nielsen, L. R., Oliveira, A. R. S. and Christensen, J. P. 2017. Evaluation of the Danish footpad lesion surveillance in conventional and organic broilers: misclassification of scoring. *Poultry Science* 96(7), 2018–28. doi:10.3382/ps/pex024.

Lynch, M., Thorp, B. H. and Whitehead, C. C. 1992. Avian tibial dyschondroplasia as a cause of bone deformity. *Avian Pathology* 21(2), 275–85. doi:10.1080/03079459208418842.

Manangi, M. K., Vazquez-Anon, M., Richards, J. D., Carter, S., Buresh, R. E. and Christensen, K. D. 2012. Impact of feeding lower levels of chelated trace minerals versus industry levels of inorganic trace minerals on broiler performance, yield, footpad health, and litter mineral concentration. *Journal of Applied Poultry Research* 21(4), 881–90. doi:10.3382/japr.2012-00531.

Mandal, R. K., Jiang, T. S., Al-Rubaye, A. A., Rhoads, D. D., Wideman, R. F., Zhao, J., Pevzner, I. and Kwon, Y. M. 2016. An investigation into blood microbiota and its potential association with bacterial chondronecrosis with osteomyelitis (BCO) in broilers. *Scientific Reports* 6, 25882. doi:10.1038/srep25882.

Manning, L., Chadd, S. A. and Baines, R. N. 2007. Water consumption in broiler chicken: a welfare indicator. *World's Poultry Science Journal* 63(1), 63–71. doi:10.1017/S0043933907001274.

Meluzzi, A., Fabbri, C., Folegatti, E. and Sirri, F. 2008a. Survey of chicken rearing conditions in Italy: effects of litter quality and stocking density on productivity, foot dermatitis and carcase injuries. *British Poultry Science* 49(3), 257–64. doi:10.1080/00071660802094156.

Meluzzi, A., Fabbri, C., Folegatti, E. and Sirri, F. 2008b. Effect of less intensive rearing conditions on litter characteristics, growth performance, carcase injuries and meat quality of broilers. *British Poultry Science* 49(5), 509–15. doi:10.1080/00071660802290424.

Muir, W. I. and Groves, P. J. 2018. Incubation and hatch management: consequences for bone mineralization in Cobb 500 meat chickens. *Animal* 12(4), 794–801. doi:10.1017/S1751731117001938.

Müller, B. R., Medeiros, H. A. S., de Sousa, R. S. and Molento, C. F. M. 2015. Chronic welfare restrictions and adrenal gland morphology in broiler chickens. *Poultry Science* 94(4), 574-8. doi:10.3382/ps/pev026.

Nääs, IdA., Paz, I. C. L. A., Baracho, M. S., Menezes, A. G., Bueno, L. G. F., Almeida, I. C. L. and Moura, D. J. 2009. Impact of lameness on broiler well-being. *Journal of Applied Poultry Research* 18(3), 432-9. doi:10.3382/japr.2008-00061.

Nääs, IdA., Paz, I. CdL. A., Baracho, MdS., Menezes, A. Gd, Lima, K. A. Od, Bueno, L. GdF., Mollo Neto, M., Carvalho, V. Cd, Almeida, I. CdL. and Souza, A. Ld 2010. Assessing locomotion deficiency in broiler chicken. *Scientia Agricola* 67(2), 129-35. doi:10.1590/S0103-90162010000200001.

Nagaraj, M., Wilson, C. A. P., Hess, J. B. and Bilgili, S. F. 2007. Effect of high-protein and all-vegetable diets on the incidence and severity of pododermatitis in broiler chickens. *Journal of Applied Poultry Research* 16(3), 304-12. doi:10.1093/japr/16.3.304.

Nicol, C. J. 2015. *The Behavioural Biology of Chickens*. CABI Publishing, Wallingford, UK.

Nicol, C. J., Caplen, G. J., Edgar, J. L. and Browne, W. J. 2009. Associations between welfare indicators and environmental choice in laying hens. *Animal Behaviour* 78(2), 413-24. doi:10.1016/j.anbehav.2009.05.016.

Nielsen, B. L., Thomsen, M. G., Sorensen, J. P. and Young, J. F. 2003. Feed and strain effects on the use of outdoor areas by broilers. *British Poultry Science* 44(2), 161-9. doi:10.1080/0007166031000088389.

Norring, M., Kaukonen, E. and Valros, A. 2016. The use of perches and platforms by broiler chickens. *Applied Animal Behaviour Science* 184, 91-6. doi:10.1016/j.applanim.2016.07.012.

Ohara, A., Oyakawa, C., Yoshihara, Y., Ninomiya, S. and Sato, S. 2015. Effect of environmental enrichment on the behavior and welfare of Japanese broilers at a commercial Farm. *Journal of Poultry Science* 52(4), 323-30. doi:10.2141/jpsa.0150034.

Olukosi, O. A., Olori, V. E., Helmbrecht, A., Lambton, S. and French, N. A. (Eds). 2019. *Poultry Feathers and Skin*. The Poultry Integument in Health and Welfare, CABI Publishing, Wallingford, UK.

Opengart, K., Bilgili, S. F., Warren, G. L., Baker, K. T., Moore, J. D. and Dougherty, S. 2018. Incidence, severity, and relationship of broiler footpad lesions and gait scores of market-age broilers raised under commercial conditions in the southeastern United States. *Journal of Applied Poultry Research* 27(3), 424-32. doi:10.3382/japr/pfy002.

Orth, M. W. and Cook, M. E. 1994. Avian tibial dyschondroplasia - a morphological and biochemical review of the growth-plate lesion and its causes. *Veterinary Pathology* 31(4), 403-4. doi:10.1177/030098589403100401.

Oso, A. O., Idowu, A. A. and Niameh, O. T. 2011. Growth response, nutrient and mineral retention, bone mineralisation and walking ability of broiler chickens fed with dietary inclusion of various unconventional mineral sources. *Journal of Animal Physiology and Animal Nutrition* 95(4), 461-7. doi:10.1111/j.1439-0396.2010.01073.x.

Oviedo-Rondón, E. O. 2019. Holistic view of intestinal health in poultry. *Animal Feed Science and Technology* 250, 1-8. doi:10.1016/j.anifeedsci.2019.01.009.

Oviedo-Rondón, E. O., Wineland, M. J., Funderburk, S., Small, J., Cutchin, H. and Mann, M. 2009a. Incubation conditions affect leg health in large, high-yield broilers. *Journal of Applied Poultry Research* 18(3), 640-6. doi:10.3382/japr.2008-00127.

Oviedo-Rondón, E. O., Wineland, M. J., Small, J., Cutchin, H., McElroy, A., Barri, A. and Martin, S. 2009b. Effect of incubation temperatures and chick transportation

conditions on bone development and leg health. *Journal of Applied Poultry Research* 18(4), 671-8. doi:10.3382/japr.2008-00135.
Oznurlu, Y., Sur, E., Ozaydin, T., Celik, I. and Uluisik, D. 2016. Histological and histochemical evaluations on the effects of high incubation temperature on the embryonic development of tibial growth plate in broiler chickens. *Microscopy Research and Technique* 79(2), 106-10. doi:10.1002/jemt.22611.
Pagazaurtundua, A. and Warriss, P. D. 2006. Levels of foot pad dermatitis in broiler chickens reared in 5 different systems. *British Poultry Science* 47(5), 529-32. doi:10.1080/00071660600963024.
Part, C. E., Edwards, P., Hajat, S. and Collins, L. M. 2016. Prevalence rates of health and welfare conditions in broiler chickens change with weather in a temperate climate. *Royal Society Open Science* 3(9), 160197. doi:10.1098/rsos.160197.
Paz, I. C. L. A., Garcia, R. G., Bernardi, R., Seno, LdO., Naas, IdA. and Caldara, F. R. 2013. Locomotor problems in broilers reared on new and re-used litter. *Italian Journal of Animal Science* 12(2). doi:10.4081/ijas.2013.e45.
Petek, M., Sonmez, G., Yildiz, H. and Baspinar, H. 2005. Effects of different management factors on broiler performance and incidence of tibial dyschondroplasia. *British Poultry Science* 46(1), 16-21. doi:10.1080/00071660400023821.
Petek, M., Cibik, R., Yildiz, H., Sonat, F. A., Gezen, S. S., Orman, A. and Aydin, C. 2010. The influence of different lighting programs, stocking densities and litter amounts on the welfare and productivity traits of a commercial broiler line. *Veterinarija ir Zootechnika* 51(73), 36-43.
Puvadolpirod, S. and Thaxton, J. P. 2000. Model of physiological stress in chickens: 4. Digestion and metabolism. *Poultry Science* 79(3), 383-90. doi:10.1093/ps/79.3.383.
Reiter, K. and Bessei, W. 2009. Effect of locomotor activity on leg disorder in fattening chicken. *Berliner Und Munchener Tierarztliche Wochenschrift* 122(7-8), 264-70.
Riddell, C. and Kong, X. M. 1992. The influence of diet on necrotic enteritis in broiler-chickens. *Avian Diseases* 36(3), 499-503. doi:10.2307/1591740.
Roberts, S. J., Cain, R. and Dawkins, M. S. 2012. Prediction of welfare outcomes for broiler chickens using Bayesian regression on continuous optical flow data. *Journal of the Royal Society Interface* 9(77), 3436-43. doi:10.1098/rsif.2012.0594.
Rushton, S. P., Humphrey, T. J., Shirley, M. D. F., Bull, S. and Jorgensen, F. 2009. Campylobacter in housed broiler chickens: a longitudinal study of risk factors. *Epidemiology and Infection* 137(8), 1099-110. doi:10.1017/S095026880800188X.
Rutten, M., Leterrier, C., Constantin, P., Reiter, K. and Bessei, W. 2002. Bone development and activity in chickens in response to reduced weight-load on legs. *Animal Research* 51(4), 327-36. doi:10.1051/animres:2002027.
Saima, M. Z. U., Jabbar, M. A., Ijaz, M. and Qadeer, M. A. 2009. Efficacy of microbial phytase at different levels on growth performance and mineral availability in broiler chicken. *Journal of Animal and Plant Sciences* 19(2), 58-62.
Sandilands, V., Brocklehurst, S., Sparks, N., Baker, L., McGovern, R., Thorp, B. and Pearson, D. 2011. Assessing leg health in chickens using a force plate and gait scoring: how many birds is enough? *Veterinary Record* 168(3), 77. doi:10.1136/vr.c5978.
Sanotra, G. S., Lund, J. D. and Vestergaard, K. S. 2002. Influence of light-dark schedules and stocking density on behaviour, risk of leg problems and occurrence of chronic fear in broilers. *British Poultry Science* 43(3), 344-54. doi:10.1080/000716601201036023611.

Sanotra, G. S., Berg, C. and Lund, J. D. 2003. A comparison between leg problems in Danish and Swedish broiler production. *Animal Welfare* 12(4), 677–83.

Sans, E., Federici, J., Dahlke, F. and Molento, C. 2014. Evaluation of Free-Range broilers using the welfare quality® protocol. *Revista Brasileira de Ciência Avícola* 16(3), 297–306. doi:10.1590/1516-635x1603297-306.

Sarica, M., Yamak, U. S. and Boz, M. A. 2014. Effect of production systems on foot pad dermatitis (FPD) levels among slow-, medium- and fast-growing broilers. *European Poultry Science* 78. doi:10.1399/eps.2014.52.

Schmidt, C. J., Persia, M. E., Feierstein, E., Kingham, B. and Saylor, W. W. 2009. Comparison of a modern broiler line and a heritage line unselected since the 1950s. *Poultry Science* 88(12), 2610–9. doi:10.3382/ps.2009-00055.

Scholz-Ahrens, K. E., Ade, P., Marten, B., Weber, P., Timm, W., Asil, Y., Gluer, C. C. and Schrezenmeir, J. 2007. Prebiotics, probiotics, and synbiotics affect mineral absorption, bone mineral content, and bone structure. *Journal of Nutrition* 137(3), 838S–46S. doi:10.1093/jn/137.3.838S.

Schwean-Lardner, K., Fancher, B. I. and Classen, H. L. 2012. Impact of daylength on behavioural output in commercial broilers. *Applied Animal Behaviour Science* 137(1–2), 43–52. doi:10.1016/j.applanim.2012.01.015.

Schwean-Lardner, K., Fancher, B. I., Gomis, S., Van Kessel, A., Dalal, S. and Classen, H. L. 2013. Effect of day length on cause of mortality, leg health, and ocular health in broilers. *Poultry Science* 92(1), 1–11. doi:10.3382/ps.2011-01967.

Sellers, H. S. 2017. Current limitations in control of viral arthritis and tenosynovitis caused by avian reoviruses in commercial poultry. *Veterinary Microbiology* 206, 152–6. doi:10.1016/j.vetmic.2016.12.014.

Senaratna, D., Samarakone, T. S. and Gunawardena, W. W. 2016. Red color light at different intensities affects the performance, behavioral activities and welfare of broilers. *Asian-Australasian Journal of Animal Sciences* 29(7), 1052–9. doi:10.5713/ajas.15.0757.

Shepherd, E. M., Fairchild, B. D. and Ritz, C. W. 2017. Alternative bedding materials and litter depth impact litter moisture and footpad dermatitis. *Journal of Applied Poultry Research* 26(4), 518–28. doi:10.3382/japr/pfx024.

Sherlock, L., McKeegan, D. E. F., Cheng, Z., Wathes, C. M. and Wathes, D. C. 2012. Effects of contact dermatitis on hepatic gene expression in broilers. *British Poultry Science* 53(4), 439–52. doi:10.1080/00071668.2012.707310.

Shim, M. Y., Karnuah, A. B., Anthony, N. B., Pesti, G. M. and Aggrey, S. E. 2012a. The effects of broiler chicken growth rate on valgus, varus, and tibial dyschondroplasia. *Poultry Science* 91(1), 62–5. doi:10.3382/ps.2011-01599.

Shim, M. Y., Karnuah, A. B., Mitchell, A. D., Anthony, N. B., Pesti, G. M. and Aggrey, S. E. 2012b. The effects of growth rate on leg morphology and tibia breaking strength, mineral density, mineral content, and bone ash in broilers. *Poultry Science* 91(8), 1790–5. doi:10.3382/ps.2011-01968.

Siegel, P. B., Gustin, S. J. and Katanbaf, M. N. 2011. Motor ability and self-selection of an analgesic drug by fast-growing chickens. *Journal of Applied Poultry Research* 20(3), 249–52. doi:10.3382/japr.2009-00118.

Simsek, U. G., Dalkilic, B., Ciftci, M., Cerci, I. H. and Bahsi, M. 2009. Effects of enriched housing design on broiler performance, welfare, chicken meat composition and serum cholesterol. *Acta Veterinaria Brno* 78(1), 67–74. doi:10.2754/avb200978010067.

Simsek, U. G., Erisir, M., Ciftci, M. and Tatli Seven, P. 2014. Effects of cage and floor housing systems on fattening performance, oxidative stress and carcass defects in broiler chicken. *Kafkas Universitesi Veteriner Fakultesi Dergisi* 20(5), 727–33.

Skrbic, Z., Pavlovski, Z., Lukic, M. and Petricevic, V. 2015. Incidence of footpad dermatitis and hock burns in broilers as affected by genotype, lighting program and litter type. *Annals of Animal Science* 15(2), 433–45. doi:10.1515/aoas-2015-0005.

Sørensen, P., Su, G. and Kestin, S. C. 2000. Effects of age and stocking density on leg weakness in broiler chickens. *Poultry Science* 79(6), 864–70. doi:10.1093/ps/79.6.864.

Sosnówka-Czajka, E., Skomorucha, I., Herbut, E. and Muchacka, R. 2007. Effect of management system and flock size on the behaviour of broiler chickens. *Annals of Animal Science* 7(2), 329–35.

Stanley, D., Hughes, R. J. and Moore, R. J. 2014. Microbiota of the chicken gastrointestinal tract: influence on health, productivity and disease. *Applied Microbiology and Biotechnology* 98(10), 4301–10. doi:10.1007/s00253-014-5646-2.

Su, G., Sorensen, P. and Kestin, S. C. 1999. Meal feeding is more effective than early feed restriction at reducing the prevalence of leg weakness in broiler chickens. *Poultry Science* 78(7), 949-55. doi:10.1093/ps/78.7.949.

Su, G., Sorensen, P. and Kestin, S. C. 2000. A note on the effects of perches and litter substrate on leg weakness in broiler chickens. *Poultry Science* 79(9), 1259–63. doi:10.1093/ps/79.9.1259.

Sun, Z. W., Yan, L., G, Y. Y., Zhao, J. P., Lin, H. and Guo, Y. M. 2013. Increasing dietary vitamin D-3 improves the walking ability and welfare status of broiler chickens reared at high stocking densities. *Poultry Science* 92(12), 3071–9. doi:10.3382/ps.2013-03278.

Sun, Z. W., Fan, Q. H., Wang, X. X., Guo, Y. M., Wang, H. J. and Dong, X. 2017. High dietary biotin levels affect the footpad and hock health of broiler chickens reared at different stocking densities and litter conditions. *Journal of Animal Physiology and Animal Nutrition* 101(3), 521–30. doi:10.1111/jpn.12465.

Swiatkiewicz, S., Arczewska-Wlosek, A. and Jozefiak, D. 2017. The nutrition of poultry as a factor affecting litter quality and foot pad dermatitis – an updated review. *Journal of Animal Physiology and Animal Nutrition* 101(5), e14–20. doi:10.1111/jpn.12630.

Tahamtani, F. M., Hinrichsen, L. K. and Riber, A. B. 2018. Welfare assessment of conventional and organic broilers in Denmark, with emphasis on leg health. *Veterinary Record* 183(6), 192. doi:10.1136/vr.104817.

Tickle, P. G., Hutchinson, J. R. and Codd, J. R. 2018. Energy allocation and behaviour in the growing broiler chicken. *Scientific Reports* 8(1), 4562. doi:10.1038/s41598-018-22604-2.

Toppel, K., Kaufmann, F., Schon, H., Gauly, M. and Andersson, R. 2019. Effect of pH-lowering litter amendment on animal-based welfare indicators and litter quality in a European commercial broiler husbandry. *Poultry Science* 98(3), 1181–9. doi:10.3382/ps/pey489.

Tullo, E., Fontana, I., Fernandez, A. P., Vranken, E., Norton, T., Berckmans, D. and Guarino, M. 2017. Association between environmental predisposing risk factors and leg disorders in broiler chickens. *Journal of Animal Science* 95(4), 1512–20. doi:10.2527/jas.2016.1257.

Tuyttens, F., Heyndrickx, M., De Boeck, M., Moreels, A., Van Nuffel, A., Van Poucke, E., Van Coillie, E., Van Dongen, S. and Lens, L. 2008. Broiler chicken health, welfare and

fluctuating asymmetry in organic versus conventional production systems. *Livestock Science* 113(2-3), 123-32. doi:10.1016/j.livsci.2007.02.019.
van der Pol, C. W., Molenaar, R., Buitink, C. J., van Roovert-Reijrink, I. A. M., Maatjens, C. M., van den Brand, H. and Kemp, B. 2015. Lighting schedule and dimming period in early life: consequences for broiler chicken leg bone development. *Poultry Science* 94(12), 2980-8. doi:10.3382/ps/pev276.
van der Pol, C. W., van Roovert-Reijrink, I. A. M., Aalbers, G., Kemp, B. and van den Brand, H. 2017. Incubation lighting schedules and their interaction with matched or mismatched post hatch lighting schedules: effects on broiler bone development and leg health at slaughter age. *Research in Veterinary Science* 114, 416-22. doi:10.1016/j.rvsc.2017.07.013.
van der Pol, C. W., van Roovert-Reijrink, I. A. M., Maatjens, C. M., Gussekloo, S. W. S., Kranenbarg, S., Wijnen, J., Pieters, R. P. M., Schipper, H., Kemp, B. and van den Brand, H. 2019. Light-dark rhythms during incubation of broiler chicken embryos and their effects on embryonic and post hatch leg bone development. *PLoS ONE* 14(1), e0210886. doi:10.1371/journal.pone.0210886.
Van Hertem, T., Norton, T., Berckmans, D. and Vranken, E. 2018. Predicting broiler gait scores from activity monitoring and flock data. *Biosystems Engineering* 173, 93-102. doi:10.1016/j.biosystemseng.2018.07.002.
Vargas-Galicia, A. J., Sosa-Montes, E., Rodriguez-Ortega, L. T., Pro-Martinez, A., Ruiz-Feria, C. A., González-Cerón, F., Gallegos-Sánchez, J., Arreola-Enríquez, J. and Bautista-Ortegac, J. 2017. Effect of litter material and stocking density on bone and tendon strength, and productive performance in broilers. *Canadian Journal of Animal Science* 97(4), 673-82.
Vasdal, G., Vas, J., Newberry, R. C. and Moe, R. O. 2019. Effects of environmental enrichment on activity and lameness in commercial broiler production. *Journal of Applied Animal Welfare Science* 22(2), 197-205. doi:10.1080/10888705.2018.1456339.
Ventura, B. A., Siewerdt, F. and Estevez, I. 2010. Effects of barrier perches and density on broiler leg health, fear, and performance. *Poultry Science* 89(8), 1574-83. doi:10.3382/ps.2009-00576.
Ventura, B. A., Siewerdt, F. and Estevez, I. 2012. Access to barrier perches improves behavior repertoire in broilers. *PLoS ONE* 7(1), e29826. doi:10.1371/journal.pone.0029826.
Villagrá, A., Olivas, I., Benitez, V. and Lainez, M. 2011. Evaluation of sludge from paper recycling as bedding material for broilers. *Poultry Science* 90(5), 953-7. doi:10.3382/ps.2010-00935.
Villarroel, M., Francisco, I., Ibanez, M. A., Novoa, M., Martinez-Guijarro, P., Méndez, J. and De Blas, C. 2018. Rearing, bird type and pre-slaughter transport conditions of broilers II. Effect on foot-pad dermatitis and carcass quality. *Spanish Journal of Agricultural Research* 16(2), e0504. Available at: http://revistas.inia.es/index.php/sjar/article/view/12015/4038 (accessed on 19 June 2019). doi:10.5424/sjar/2018162-12015.
Waldenstedt, L. 2006. Nutritional factors of importance for optimal leg health in broilers: a review. *Animal Feed Science and Technology* 126(3-4), 291-307. doi:10.1016/j.anifeedsci.2005.08.008.
Weeks, C. A. and Butterworth, A. 2004. *Measuring and Auditing Broiler Welfare* (vol. 1). CABI Publishing, Wallingford, UK.

Weeks, C. A., Danbury, T. D., Davies, H. C., Hunt, P. and Kestin, S. C. 2000. The behaviour of broiler chickens and its modification by lameness. *Applied Animal Behaviour Science* 67(1-2), 111-25. doi:10.1016/s0168-1591(99)00102-1.

Weeks, C. A., Knowles, T. G., Gordon, R. G., Kerr, A. E., Peyton, S. T. and Tilbrook, N. T. 2002. New method for objectively assessing lameness in broiler chickens. *Veterinary Record* 151(25), 762-4.

Weimer, S. L., Wideman, R. F., Scanes, C. G., Mauromoustakos, A., Christensen, K. D. and Vizzier-Thaxton, Y. 2019. The utility of infrared thermography for evaluating lameness attributable to bacterial chondronecrosis with osteomyelitis. *Poultry Science* 98(4), 1575-88. doi:10.3382/ps/pey538.

Welfare Quality. 2009. Welfare Quality assessment protocol for poultry (broilers, laying hens). Welfare Quality Consortium, Lelystad, the Netherlands. Available at: http://www.welfarequalitynetwork.net/downloadattachment/45627/21652/Poultry%20Protocol.pdf (accessed on 19 June 2019).

Whitehead, C. C., McCormack, H. A., McTeir, L. and Fleming, R. H. 2004. High vitamin D-3 requirements in broilers for bone quality and prevention of tibial dyschondroplasia and interactions with dietary calcium, available phosphorus and vitamin A. *British Poultry Science* 45(3), 425-36. doi:10.1080/00071660410001730941.

Wideman, R. F. 2016. Bacterial chondronecrosis with osteomyelitis and lameness in broilers: a review. *Poultry Science* 95(2), 325-44. doi:10.3382/ps/pev320.

Wideman, R. F. and Prisby, R. D. 2013. Bone circulatory disturbances in the development of spontaneous bacterial chondronecrosis with osteomyelitis: a translational model for the pathogenesis of femoral head necrosis. *Frontiers in Endocrinology* 3, 183. doi:10.3389/fendo.2012.00183.

Wideman, R. F., Hamal, K. R., Stark, J. M., Blankenship, J., Lester, H., Mitchell, K. N., Lorenzoni, G. and Pevzner, I. 2012. A wire-flooring model for inducing lameness in broilers: evaluation of probiotics as a prophylactic treatment. *Poultry Science* 91(4), 870-83. doi:10.3382/ps.2011-01907.

Wideman, R. F., Al-Rubaye, A., Reynolds, D., Yoho, D., Lester, H., Spencer, C., Hughes, J. D. and Pevzner, I. Y. 2014. Bacterial chondronecrosis with osteomyelitis in broilers: influence of sires and straight-run versus sex-separate rearing. *Poultry Science* 93(7), 1675-87. doi:10.3382/ps.2014-03912.

Wideman, R. F., Blankenship, J., Pevzner, I. Y. and Turner, B. J. 2015. Efficacy of 25-OH vitamin D-3 prophylactic administration for reducing lameness in broilers grown on wire flooring. *Poultry Science* 94(8), 1821-7. doi:10.3382/ps/pev160.

Wijesurendra, D. S., Chamings, A. N., Bushell, R. N., Rourke, D. O., Stevenson, M., Marenda, M. S., Noormohammadi, A. H. and Stent, A. 2017. Pathological and microbiological investigations into cases of bacterial chondronecrosis and osteomyelitis in broiler poultry. *Avian Pathology* 46(6), 683-94. doi:10.1080/03079457.2017.1349872.

Wijtten, P. J. A., Hangoor, E., Sparla, J. K. and Verstegen, M. W. A. 2010. Dietary amino acid levels and feed restriction affect small intestinal development, mortality, and weight gain of male broilers. *Poultry Science* 89(7), 1424-39. doi:10.3382/ps.2009-00626.

Xavier, D. B., Broom, D. M., McManus, C. M. P., Torres, C. and Bernal, F. E. M. 2010. Number of flocks on the same litter and carcase condemnations due to cellulitis, arthritis and contact foot-pad dermatitis in broilers. *British Poultry Science* 51(5), 586-91. doi:10.1080/00071668.2010.508487.

Yalçin, S., Molayoglu, H. B., Baka, M., Genin, O. and Pines, M. 2007. Effect of temperature during the incubation period on tibial growth plate chondrocyte differentiation and the incidence of tibial dyschondroplasia. *Poultry Science* 86(8), 1772–83. doi:10.1093/ps/86.8.1772.

Yamak, U. S., Sarica, M., Boz, M. A. and Ucar, A. 2016. Effect of reusing litter on broiler performance, foot-pad dermatitis and litter quality in chickens with different growth rates. *Kafkas Universitesi Veteriner Fakultesi Dergisi* 22(1), 85–91.

Yan, F. F., Wang, W. C. and Cheng, H. W. 2018. Bacillus subtilis based probiotic improved bone mass and altered brain serotoninergic and dopaminergic systems in broiler chickens. *Journal of Functional Foods* 49, 501–9. doi:10.1016/j.jff.2018.09.017.

Yan, F. F., Mohammed, A. A., Murugesan, G. R. and Cheng, H. W. 2019. Effects of a dietary synbiotic inclusion on bone health in broilers subjected to cyclic heat stress episodes. *Poultry Science* 98(3), 1083–9. doi:10.3382/ps/pey508.

Yang, H. M., Xing, H., Wang, Z. Y., Xia, J. L., Wan, Y., Hou, B. and Zhang, J. 2015. Effects of intermittent lighting on broiler growth performance, slaughter performance, serum biochemical parameters and tibia parameters. *Italian Journal of Animal Science* 14(4), 4143. doi:10.4081/ijas.2015.4143.

Yildiz, H., Gunes, N., Cengiz, S., Ozcan, R., Petek, M., Yilmaz, B. and Arican, I. 2009. Effects of ascorbic acid and lighting schedule on tibiotarsus strength and bone characteristics in broilers. *Archiv fur Tierzucht* 52(4), 432–44.

Yildiz, A., Yildiz, K. and Apaydin, B. 2014. The effect of vermiculite as litter material on some health and stress parameters in broilers. *Kafkas Universitesi Veteriner Fakultesi Dergisi* 20(1), 129–34.

Zhao, J., Shirley, R. B., Vazquez-Anon, M., Dibner, J. J., Richards, J. D., Fisher, P., Hampton, T., Christensen, K. D., Allard, J. P. and Giesen, A. F. 2010. Effects of chelated trace minerals on growth performance, breast meat yield, and footpad health in commercial meat broilers. *Journal of Applied Poultry Research* 19(4), 365–72. doi:10.3382/japr.2009-00020.

Zhao, Z. G., Li, J. H., Li, X. and Bao, J. 2014. Effects of housing systems on behaviour, performance and welfare of fast-growing broilers. *Asian-Australasian Journal of Animal Sciences* 27(1), 140–6. doi:10.5713/ajas.2013.13167.

Zikic, D., Djukic-Stojcic, M., Bjedov, S., Peric, L., Stojanovic, S. and Uscebrka, G. 2017. Effect of litter on development and severity of foot pad dermatitis and behavior of broiler chickens. *Revista Brasileira de Ciência Avícola* 19(2), 247–54. doi:10.1590/1806-9061-2016-0396.

CPSIA information can be obtained
at www.ICGtesting.com
Printed in the USA
BVHW011727110221
599709BV00016B/555